CONTENTS

Metals

Drinks

Warmth

Clothing

Food

Transporting chemicals

Plastics

Growing food

Food processing

Minerals

Keeping clean

CHEMISTRY
The Salters' Approach

GRAHAM HILL
JOHN HOLMAN
JOHN LAZONBY
JOHN RAFFAN
DAVID WADDINGTON

HEINEMANN
EDUCATIONAL

540

To the teachers and students who helped us and took part in the trials of the
Salters' Course
and to Francesca Garforth for her inspiration
and John Montgomery for his encouragement.

Heinemann Educational Books Ltd
Halley Court, Jordan Hill, Oxford OX2 8EJ

Oxford   London   Edinburgh
Madrid   Athens   Bologna
Melbourne   Sydney   Auckland
Ibadan   Nairobi   Gaborone   Harare
Kingston   Portsmouth (NH)   Singapore

ISBN 0 435 64000 3

© University of York Science Education Group, 1989

First published 1989
Reprinted 1990

Buildings

Making and using electricity

Burning and bonding

Fighting disease

Energy today and tomorrow

CHEMISTRY:
THE SALTERS' APPROACH

About the approach

The Salters' approach is an entirely new way of teaching and learning chemistry for GCSE. It is an approach which goes further than simply putting chemistry into context.
◆ Each topic is based on aspects of everyday life.
◆ Chemical concepts and explanations arise naturally from the study of these everyday situations.
The approach was developed by a large and experienced group of school and university teachers. They felt that the Salters' approach needed an entirely new kind of textbook.

About this book

This book has been written to accompany the units of the Salters' Course. It contains sixteen chapters, one for each unit, and each is divided into five key sections.

Introducing pages set the scene of the topic and raises some questions about what is in it. You should read this before starting the chapter.

The **Looking at** sections are short pieces on subjects relating to the theme of the chapter. You are not expected to read or learn about all of them. You can choose the ones you think are interesting. Or your teacher may ask you to read particular ones. These sections include questions to make you think more deeply about the main points.

In brief sections give you a straightforward summary of what you need to know to understand the topic.

Thinking about pages explain the key chemical ideas that arise from each topic. When you read the *In brief* section, for example, you might realise that you need to look at a particular *Thinking about* section, or your teacher might ask you to study one. When you read this section for the first time you may decide to leave out the *Taking it further* boxes.

Things to do is a bank of activities for you to try and they include:
◆ *Activities to try*: investigations to do in the laboratory or at home
◆ *Things to find out*: questions to research from other books
◆ *Things to write about*
◆ *Points to discuss*: these are best discussed in small groups of three, four or five
◆ *Questions to answer*: these will help you when it comes to exams

In writing this book we set ourselves the task of showing you how chemistry affects your lives, and then helping you to understand the chemical ideas which explain these effects. In doing this, we had to learn a lot ourselves. We enjoyed writing the book; we hope you will enjoy using it.

Graham Hill, John Holman, John Lazonby, John Raffan, David Waddington

Figure 1 *The metal alloy which is used for the outer casing of this spaceship is strong, has a low density and is heat resistant.*

Introducing metals

You only have to look around you to see how important metals are in your life. There is a vast array of different metals which are used in different ways for different purposes.

People have not always had so many metals available to them. Most metals have to be extracted from rocks. You will find out more about how this is done later on in the course. Copper was the first metal to be used as it was the easiest to extract. Later people began to use bronze and iron. Aluminium, which is an important metal, was not produced commercially until 1886.

Now it is possible to obtain specially prepared mixtures of metals called **alloys** which have suitable properties for making everything from a spoon to a spaceship.

Figure 2
Choosing the right metal for the right job is vital!

Explain what is wrong with the choices made by the people in figure 2.

In this chapter you will see how

◆ metals have some similarities which enable us to know that an object is made from metal rather than any other material,
◆ metals also have many differences –

in particular some metals corrode more easily than others,
◆ an understanding of the chemical reactions involved in corrosion can help to prevent it.

1 Poisonous metals

Figure 3 *Wealthy Greeks and Romans often used slaves to taste all their food and drink to see if it was safe to eat - a wise precaution!*

The ancient Greeks and Romans were in the habit of poisoning each other. Aristotle was executed by poisoning in 399 BC and it is thought that Agrippina poisoned Claudius with arsenic so that Nero could become emperor and then Nero himself poisoned Britannicus who was the son of Claudius. Clearly poisoning was a required political skill in those days. Because of stories such as these, there is a temptation to define a poison as a substance which kills someone but this is a very unscientific definition. Many substances which are essential in your diet can be poisonous if you take them in excess. Other more poisonous substances may be quite safe in very small quantities. So describing substances as either poisonous or safe is not very helpful. It is much more useful to know how poisonous substances are compared with each other.

Table 1 shows the comparative **toxicity** (how poisonous they are) of some substances.

The comparison is based on animal tests. The dose per kg of body weight which kills 50 per cent of the tested animals is called the Lethal Dose 50 per cent, or the LD50.

So the lower the LD50, the more lethal the substance. These values are probably useful for rough comparisons but we cannot be sure that the effect on humans is the same

Table 1

Substance	LD50/mg per kg	Toxicity
Alcohol	10 000	slightly toxic
Sodium chloride	4000	moderately toxic
DDT (insecticide)	100	very toxic
Nicotine	1	super toxic
Dioxin (impurity in a herbicide)	0.001	

as the effect on mice. Obviously we should even be concerned about doses which are many times smaller than the LD50.

In addition to the substances mentioned in table 1 there are a number of metals, sometimes called heavy metals, which are particularly **toxic**. Mercury, cadmium and lead are toxic heavy metals.
The poisonous nature of lead is probably due to its similarity to metals such as calcium which are essential to your health. For example, lead can replace the calcium in your bones and so stay in your body. So the concentration of lead in your body can continue to increase. For this reason lead is said to have a **cumulative effect**.

Lead poisoning can be treated with injections of a substance which will react more strongly with the lead than the compounds which are holding it in the body. This results in the lead being excreted from the body.

Figure 4 *Lead water pipes have been used since Roman times, and can be a cause of lead poisoning.*

doses of lead over a period of time. There is evidence that young children may be particularly affected by relatively low doses of lead.

The major source of these small doses of lead is likely to be the **lead in petrol**. It is added to petrol in the form of a compound called **tetraethyl lead**. This is added to prevent the engine 'knocking' (which is caused by the mixture of air and petrol vapour exploding before the spark ignites it). The lead compound increases the efficiency of the engine but unfortunately the lead compounds pass out of the engine in the exhaust fumes (figure 5).

People living in urban areas are likely to be exposed to the greatest risk from these lead compounds. More and more countries are starting to use lead-free petrol (figure 6).

Figure 5 *Busy urban roads like this can cause dangerous levels of lead to build up in the surrounding area.*

Causes of lead poisoning

Lead poisoning occurred in Roman times. If they drank slightly acidic wine from their lead goblets they would inevitably absorb some lead. The Romans also used **lead water pipes** (figure 4) and in fact their word for lead – *plumbum* has given us the word plumbing and also the symbol for lead, Pb.

Figure 6 *Lead-free petrol is now becoming more readily available and many cars now use it.*

Besides the continued use of lead pipes, the other major source of lead poisoning in the first half of this century was the production and use of **lead-based paints**. Lead is also used in the production of **car batteries**, so during the early days of their production the workers frequently developed lead poisoning. The use of lead paints is very much restricted and controlled now. Once the scientific evidence was collected about the harmful effects of lead, the working conditions in factories were made safe too.

Except for accidents, severe cases of lead poisoning are rare. However, there is concern about the toxic effects of small

1 What sources of lead poisoning might exist in old houses?
2 Why is it particularly important that toys and cots should not be painted with lead-based paints?
3 When lead poisoning was first investigated in factories its presence was only recognised when black marks appeared in the gums of the workers.
Techniques are now available which can measure concentrations of lead in blood down to 0.00001g in 100g. Detecting lead at such low levels in the blood of people living in a city area may not be accepted as a sufficient reason for banning lead in petrol. What sort of arguments and evidence might be used for the case against banning lead in petrol?

2 Using one metal to save another

Iron is chosen to make so many objects because it is relatively cheap and it is strong (figure 7).

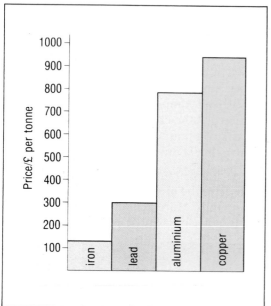

Figure 7 *Comparing the costs of some metals. Gold costs about £10 000 000 per tonne. Using the same scale as in the bar chart above, how high would the bar for gold be?*

But iron does have a serious disadvantage. It reacts very readily with air and water. This corrosion of iron is commonly called **rusting**. Other metals also **corrode** but the term rusting is used only for the corrosion of iron.

Other metals are often used to protect iron. The tin on the food can (figure 8) and the zinc on the dustbin (figure 9) protect the iron by keeping air and water away from it. Iron can only rust if it comes into contact with air *and* water.

The lumps of magnesium are obviously not covering the whole of the ship in figure 10 and so they must be protecting the iron in another way. Magnesium is more reactive than the iron. This means that the magnesium is corroded in preference to the iron. The magnesium is sacrificed to save the iron.

The magnesium lumps need to be replaced at regular intervals, but this is cheaper than replacing the ship!

Figure 8 *Tin cans are not just made of tin. They are made of iron which has been covered with a thin layer of tin. This makes them safe to put food in. Why?*

Figure 9 *A zinc dustbin is not just made of zinc. It is made of iron coated with zinc (called galvanised iron). Why is zinc used?*

Figure 10 *Lumps of magnesium are welded to the iron hulls of ships to protect them from rusting. You can see six magnesium bars in this picture. How do they protect the ship?*

Figure 11 compares the reactivity of different metals. This comparison is often called the **reactivity series**.

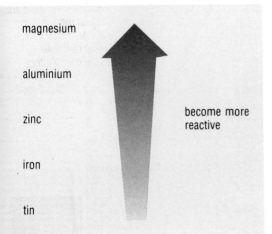

magnesium

aluminium

zinc — become more reactive

iron

tin

Figure 11   *Some metals in the reactivity series*

Aluminium is well above iron in the reactivity series and yet it does not appear to need special protection from corrosion. The secret is a very thin layer of aluminium oxide which forms on the surface of a freshly cut piece of aluminium when it is left in air (figure 12).

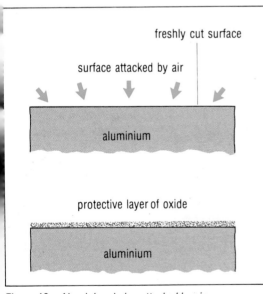

freshly cut surface

surface attacked by air

aluminium

protective layer of oxide

aluminium

Figure 12   *Aluminium being attacked by air*

This layer of oxide sticks to the aluminium and prevents any more air attacking the rest of the aluminium. The aluminium has formed its own protective layer.

This layer can be artificially thickened using a process called **anodising**. You will find out more about this later.

1 Suggest reasons why strips of magnesium are used to protect the iron hulls of ships but not to protect iron gutters and drain pipes on a house?
2 The metal objects in figures 13 and 14 are protected from corrosion by thin layers of metal which also make them attractive. What metals do you think are being used and why is it a different metal in each case?

Figure 13

3 When metals corrode they usually change into the metal oxide. In the case of aluminium this oxide then protects the rest of the aluminium. Suggest a possible reason why iron oxide does not protect iron in the same way.

3 Why are metals hard?

It is possible to make solid objects from wood, plastic, pottery or metal. They are all **hard** in the sense that if you are hit by an object made from one of these materials it hurts. Obviously the one you would least like to be hit with is the metal object, but bumping into objects is not a very scientific way of comparing the materials they are made of!

But there are several ways of defining the meaning of hard.

One meaning is to do with the hardness of the surface – which material is most difficult to scratch or cut?

Another meaning is strength (figure 15) – if one end of the object is fixed and the other end is pulled, would it stretch or break?

A third meaning might be to do with what happens if you try to bend the material – does it bend or does it snap?

If you consider those metals such as aluminium, copper, iron and tin which are used to make objects, then in general these metals compared to wood, plastic or pottery:

◆ have surfaces which are more difficult to scratch or cut,
◆ are stronger when tested by pulling,
◆ are more bendy and are less likely to snap when bent.

But why are some metals harder than other materials? You will get a clue by looking closely at their surfaces (figure 16).

Each grain is a single **crystal** of the metal. When a metal is first extracted from its **ore** it is a liquid. As the liquid cools it starts to solidify as tiny crystals form in different parts of the liquid. As the metal continues to cool the crystals grow until they meet the other crystals around them and then the whole metal is a solid mass of tiny crystals.

The hardness of the metal is linked to its crystalline form and how strongly the crystals are held together.

But some metals are much harder than others and even different pieces of the same metal can have different hardnesses (figure 17).

Figure 15 The steel cables used by cranes like this one are strong enough to lift and hold very heavy objects.

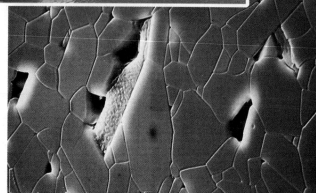

Figure 16 This photograph shows the magnified surface of aluminium. It shows that it is made up of tiny grains tightly packed together.

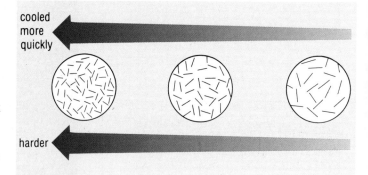

cooled more quickly

harder

Figure 17 Cooling liquid metals

1 How do you think the hardness of the metal is influenced by the number of crystals and their size?
2 Draw a table which lists the three properties of metals mentioned in this section. Include additional columns for wood, plastic and pottery. Fill in next to each property how you think these other three substances differ from metals.

1 Metals are used in the construction of buildings, bicycles, cars, trains, aeroplanes and countless everyday objects used in the home, at work and in leisure pursuits. Metals are used when their properties suit the task and the cost of producing them matches what people are willing to pay.

2 Metals have particular physical properties which make them different from other materials such as plastic, wood or pottery.

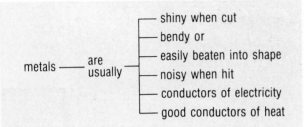

Figure 18 *Physical properties of metals*

Not all metals are good examples of all of these properties, but usually each one shows enough of them for you to know that it is a metal.

3 Although metals have a lot of similarities, they also have a lot of differences. These differences are important when deciding which metal to use for which purpose. For example, some metals are stronger than others, some are heavier than others and some are more resistant to attack by air.

4 When a metal is attacked by air, it has been **corroded**. Corrosion is a **chemical change** because the metal reacts with the oxygen in the air (and sometimes water) to form a new substance.

$$\text{metal} + \text{oxygen} \rightarrow \text{new substance}$$
$$\text{(and water)}$$

In the case of iron, the new substance formed is iron oxide, called rust.

5 Some metals corrode more readily than others. They are more reactive. The order of reactivity of some of the metals is shown below. It is called the metal **reactivity series** (figure 19).

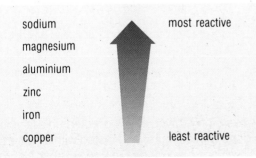

sodium — most reactive
magnesium
aluminium
zinc
iron
copper — least reactive

Figure 19 *Some metals in the reactivity series*

6 To prevent the corrosion of a metal you can:
   ◆ keep air and moisture away from the metal,
   ◆ change the properties of the metal by alloying it,
   ◆ fix small pieces of a more reactive metal to its surface.

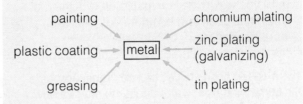

Ways of preventing corrosion by covering the metal

painting   chromium plating

plastic coating → metal ← zinc plating (galvanizing)

greasing   tin plating

7 An **alloy** is a mixture of two or more metals (figure 20). It is made by mixing the metals together as liquids and then allowing them to cool and solidify.

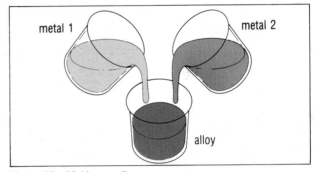

Figure 20 *Making an alloy*

Alloys have different properties from any of the metals they contain.

8 A pure metal is an element which means that it cannot be broken down into anything simpler. For example, pure copper is just copper and nothing else.

In the same way pure non-metals such as oxygen and carbon are elements because they cannot be broken down into anything simpler.

Each element can be represented by a **symbol**. The symbol is one or two letters from the name of the element. For example, magnesium is Mg, iron is Fe (which is based on the Latin name for iron *ferrum*) and oxygen is O. The symbols are a type of shorthand for the elements; they are understood everywhere in the world.

The product of corrosion, such as rust, is a **compound** because it contains more than one element (iron and oxygen).

1 What happens when iron rusts?

If you investigated what happens when iron rusts you would find that:

◆ iron nails will only rust when both water and air are present,
◆ the iron seems to be 'eaten away' by the rusting, but if you weigh the iron that is left with the rust which has been formed, there has been an increase in mass,
◆ rust looks very different from the original iron.

These observations tell us that iron must combine with something from water and air to make a new substance which we call rust (figure 21).

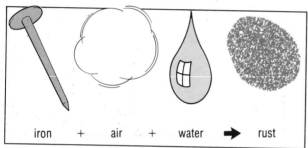

iron + air + water ➡ rust

Figure 21

Figure 24 *Gold objects such as this mask from an Egyptian tomb will stay shiny for ever.*

2 How do the reactivities of metals differ?

Look at figures 22 – 24.

Figure 22 *Sodium is so reactive that it must be kept away from air and so it is stored under oil.*

Figure 23
Iron which is not protected from a damp atmosphere will rust.

You can see from these photographs that sodium is more reactive than iron, whereas gold is much less reactive than either of the other two metals.

Similar differences are found if the reactions of metals with water are investigated (table 2).

Table 2

Sodium	reacts violently with water and gives off a gas
Calcium	reacts steadily with water and gives off a gas
Magnesium	reacts very slowly with water and gives off a gas
Iron	reacts with water and air after a day or so to form rust
Copper	does not react with water

If the gas which is given off is collected (this is most easily done with calcium) and tested with a lighted splint, it will pop or explode. This shows that it is hydrogen.

Taking it further

When sodium or calcium or magnesium react with water, the gas given off is hydrogen. The solution which is left turns litmus blue showing that it is an **alkali**. Alkalis are metal hydroxides, so the reaction of, for example, calcium with water can be summarised by the word equation:

calcium + water → calcium + hydrogen
hydroxide

Similar equations can be used to summarise the reactions of sodium and magnesium with water.

odium   most
alcium   reactive
nagnesium
luminium
on
nc
opper
old

Figure 25 *Building up the reactivity series*

Evidence such as how metals react with air and water helps us to build up the reactivity series. Figure 25 shows a more complete series.

Hydrogen and oxygen are both elements. Pure water (H_2O) is not an element it is a compound in which hydrogen and oxygen are joined together. It is possible, although not so easy, to get the hydrogen and oxygen out of water, but it is not possible to get anything out of hydrogen or to get anything out of oxygen.

You cannot tell by just looking at a substance that it is an element. It is just by experience that you learn which substances are elements and which are not.

Everything around you is made up of elements. You are, the chair you are sitting on is, the air you breathe, the food you eat – in fact the whole universe is made up of elements and combinations of elements.

However there is only a limited number of elements. The sun, the moon, the planets, Halley's comet; none of them can contain any elements which are not found on the earth (figures 28 and 29).

3   What is a chemical element?

A pure metal such as copper is an element – it contains nothing but copper. Brass is not an element because it contains a mixture of copper and zinc – it is an alloy of two elements.

▲ **Figure 26**
*Copper is an element (**figure 26**) but brass is made from a mixture of copper and zinc (**figure 27**). Brass is an alloy.*

Figure 27 ▶

Figure 28

Figure 29
Moon rock is made from elements which are all found on the earth too.

4 How do chemists represent elements?

Chemical names can be very long. They also vary from one language to another. This could cause considerable confusion.

Figure 30 *What are in these bottles?*

So chemists use a system of shorthand for the elements. They write a symbol which is either one or two letters instead of writing out the whole name. For example,

O stands for oxygen
and Mg stands for magnesium.

These symbols are part of an international language. So a chemist, or chemistry student, anywhere in the world will know that Mg is magnesium. This is just like a pianist in any country being able to recognise the way Chopin wrote down his music or a guitarist in any country recognising the chord symbols used to represent a tune.

An element is represented by the first letter of its name, or if there are two or more elements starting with a particular letter, then another letter from its name is also used:

Carbon is represented	by C
Calcium	by Ca
Cobalt	by Co
Chlorine	by Cl
Chromium	by Cr

Sometimes the symbol is derived from the Latin name for the element:

Sodium (*Natrium*)	by Na
Copper (*Cuprum*)	by Cu
Iron (*Ferrum*)	by Fe

5 How can the properties of metals be modified?

Electricians or electronics engineers need to be able to join components to copper wires. The joint must conduct electricity and so it must be made of metal. When the metal is heated gently it should melt without melting the copper or damaging the components and when it cools it should form a solid metal joint.

Solder is an alloy of lead and tin and has the ideal properties for joining electrical components. It melts at a lower temperature than either of the pure metals and so doesn't damage the components (figure 31).

Figure 31 *Solder (an alloy of lead and tin) can be melted with a hot soldering iron and used to join electrical components without damaging them.*

Most metals in everyday use are in fact alloys. By adding another metal, or occasionally a non-metal, to the pure metal its properties can often be changed to make it suitable for different purposes. Figures 32 and 33 show some common alloys.

◀ **Figure 32** *This steel scalpel is made from an alloy of iron and carbon. It is stronger than pure iron.*

Figure 33 *This aeroplane is made of a light and strong titanium alloy.* ▼

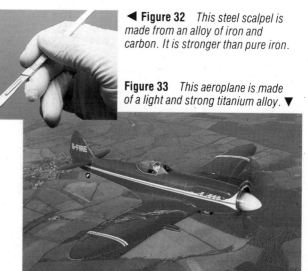

Things to do

Things to try out

1 Identify five objects in your home which you think are made of metal. Without damaging the objects carry out tests on them to decide if they are definitely made of metal. Draw a table with the names of the objects in one column. In a second column list the properties which have helped you to decide that they are made of a metal.

2 A food manufacturer produces tinned vegetables and tinned fruit. The vegetables are stored in salt solution and the fruit in sugar solution. Design and carry out your own investigation into the effect of these two solutions on dented tin cans.

Things to find out

3 With the help of older members of your family identify three objects which used to be made of metal and are now made of plastic. By discussing these objects with your family try to decide whether metal is no longer used for each of these objects because it is too expensive or because the plastic has more suitable properties.

4 Whenever a way of extracting a metal from its ore was discovered the metal became available for making useful objects. By consulting other books find out when some of the more important metals were first extracted and construct a time chart, possibly starting with copper. Metals which could be included are: aluminium, chromium, iron, lead, tin, tungsten and zinc.

Things to write about

5 Figure 34 shows the composition of the earth's crust. Copper is one of the metals in the 'other elements' section. It is thought that about 0.007 per cent of the earth's crust is copper.

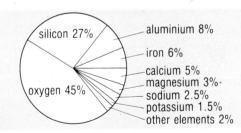

Figure 34 *Components of the earth's crust*

Most copper ores contain no more than 1 per cent of copper.

The consumption of copper metal in the USA works out at about 8 kg per head of population each year. In India it is about 0.1 kg per head of population.

The population of the USA is about 240 million and the population of India is about 770 million.

Write a magazine article, with an eye-catching title, which discusses what you think is important about these figures.

6 There have been particular incidents of mercury poisoning in the sea off Japan and in the Great Lakes of America. Using other books find out what you can about the source of the mercury poisoning and how it came to affect people.

Making decisions

Table 3

Metal (scale 1 – 3) (1 = most)	A	B	C	D
Strength	1	3	2	1
Flexibility	3	1	2	2
Corrodability	2	2	2	3
Density	medium	low	low	medium
Cost	medium	cheap	cheap	expensive

7 Which of the metals, A to D, would you choose to make:

a) a pressurised can for deodorant spray,
b) a fret for a guitar,
c) a toothpaste tube,
d) the metal part of a diamond stylus?

Points to discuss

8 The average consumption of milk in the UK is 240 pints per person per year. If you assume that 80 per cent of it is delivered in bottles with aluminium tops and that the mass of a bottle top is about 0.2 g, estimate the total mass of aluminium used to make milk bottle tops per year. What are your reactions to these data? What other data do you think you need to know before you can come to a definite view on the use, conservation and recycling of aluminium?

9 A motor car exhaust pipe which is made of stainless steel will corrode much more slowly than one made from normal steel. However, it costs about twice as much as a normal exhaust pipe. How would you decide whether or not to use a stainless steel exhaust pipe?

Questions to answer

10 A Aluminium
B Copper
C Steel
D Chromium
E Sodium

Choose from the metals A to E, the one which is

a) an alloy,
b) NOT an element,
c) used to make milk bottle tops,
d) used to protect iron,
e) a pink/brown colour,
f) the most reactive with water,
g) represented by the symbol Cu,
h) corroded to form rust.

11 Look at the Periodic Table below (figure 35). Each square contains the symbol for an element. The table can be thought of as an arrangement of all of the elements which results in similar elements being near each other.

a) Nickel (Ni) is used to make coins. Locate it in the table and find and name two other metals which are used to make coins.
b) Gold (Au) is used to make jewellery. Locate it in the table and find and name two other metals which are used for the same purpose.
c) Potassium (K) reacts vigorously with cold water. Locate it in the table. Find and name another metal which reacts in the same way with water.
d) Cobalt (Co) is magnetic. Locate it in the table. Find and name another metal which is also magnetic.

12 State, with reasons, whether each of the following substances is an element, an alloy or a compound:

brass, copper, lead, water, duralium, copper(II) oxide, rust, iron, solder, magnesium oxide, oxygen.

13 Based on their reactions with oxygen it is possible to arrange the following metals in the order of reactivity indicated. The most reactive metal is at the top of the list.

sodium
magnesium
iron
copper

Explain which of the following reactions you would predict to be likely to occur and which would not be likely to occur.

magnesium + copper(II) oxide → copper + magnesium oxide

iron + sodium oxide → sodium + iron oxide

group I	group II											group III	group VI	group V	group VI	group VII	group 0
						H											He
Li	Be											B	C	N	O	F	Ne
Na	Mg											Al	Si	P	S	Cl	Ar
K	Ca	Sc	Ti	V	Cr	Mn	Fe	Co	Ni	Cu	Zn	Ga	Ge	As	Se	Br	Kr
Rb	Sr	Y	Zr	Nb	Mo	Tc	Ru	Rh	Pd	Ag	Cd	In	Sn	Sb	Te	I	Xe
Cs	Ba	La	Hf	Ta	W	Re	Os	Ir	Pt	Au	Hg	Ti	Pb	Bi	Po	At	Rn
Fr	Ra	Ac															

Figure 35 *Periodic table*

Figure 1 *The water collected from this stream will be used for drinking and cooking. The stream may also be used for washing and sanitation so there is great danger that it will become infected.*

Introducing drinks

People go on hunger strikes to make a strong protest. One of the most famous hunger strikes was by the suffragettes 70 years ago. The suffragettes wanted votes for women. Later Mahatma Gandhi went on hunger strike against the British authorities in India. Gandhi was trying to win freedom for his country. Unlike hunger strikes, we never hear of thirst strikes.

Everyday you lose two litres of water through sweating and urinating. So everyday you need to replace this water. After breathing, drinking is the most necessary activity for sustaining life. People can live for 50 to 60 days without eating but will die after five to ten days without water.

For thousands of years, people took their drinking water from the nearest river or stream. As towns and cities got larger, this became very unhealthy. All kinds of domestic and industrial waste was being dumped in the same streams from which people took their drinking water. This led to outbreaks of disease. For example, 50 000 people died from cholera in England in 1831 when the Thames and other rivers became infected. You can now take it for granted that in Britain there will always be clean and safe water coming from the taps. However, this is not so in many other countries.

There are huge numbers of people in the world who do not have their own supply of water. Some people have to walk many miles each day to collect water. Others may only have a water supply which is polluted and unclean.

◀ **Figure 2** *By building protected wells, like this one, for drinking water, the risk of infection is greatly reduced.*

In this chapter you will see how

◆ chemistry helps us to clean and purify water,
◆ chemistry helps us to produce and manufacture different kinds of drinks.

1 Think before you drink

What does the law say about alcohol?

◆ It is against the law to sell alcoholic drinks to anyone under the age of eighteen.

◆ No-one under eighteen can work in a bar.

◆ No-one under fourteen is allowed in a bar or an off-licence.

In spite of these laws, 91 per cent of seventeen year old boys and 35 per cent of seventeen year old girls admitted drinking illegally in pubs in a recent survey.

Breath test and breath test failures (1986)

Car drivers	Involved in incident	Tested	Failed	Failed as %	
				Involved	Tested
under 17	981	113	38	3.9%	33.6%
17 - 19	25 498	6892	1057	4.1%	15.3%
20 - 24	49 123	11 421	2724	5.5%	23.9%
25 - 28	31 668	6235	1590	5.0%	25.5%
29 - 34	38 375	10 622	1577	4.1%	23.8%
All ages	290 580	49 559	10 014	3.4%	20.2%

1 What percentage of all drivers involved in an incident failed a breath test?

2 Work out what percentage of 20 - 24 year olds involved in incidents were tested.

3 List two arguments for and two against random breath testing.

What happens if you drink alcohol?

Although most young people learn to drink moderately and safely as they get older, some teenagers drink too heavily. This causes serious problems at home and at school.

Young people seem to be affected by alcohol more easily than adults. Drinking begins to affect their judgement well before they reach the legal limit for drinking and driving (figure 3). At the legal limit, a young driver is five times more likely to have an accident than when he or she has not been drinking.

Be very careful about when and how much you drink. Look closely at what the Law says. Alcohol is a **strong drug**. It can affect your health and your whole life.

Figure 3 *Testing a driver's breath to see if he has consumed too much alcohol to drive.*

How much alcohol do drinks contain?

Different drinks contain different amounts of alcohol. Table 2 on page 20 shows the percentage of alcohol in different drinks. But remember the volume of the drink is important as well as the percentage of alcohol in it. Figure 4 opposite shows the normal measures sold in pubs. Each of these contain about the same amount of alcohol.

What happens if you drink alcohol?

After drinking, liquid passes quickly from the mouth to the stomach and then into the intestine (figure 5). Only small particles can pass through the lining of the stomach into the blood. Alcohol particles are small enough to do this. So you feel the effects of alcohol very soon after drinking it.

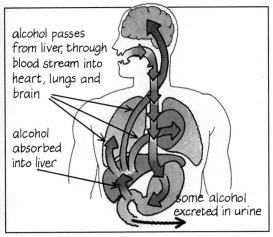

alcohol passes from liver, through blood stream into heart, lungs and brain

alcohol absorbed into liver

some alcohol excreted in urine

Figure 5   *Where the alcohol goes*

If you eat at the same time, alcohol is absorbed by the food. So it takes longer to get into your blood.

From the small intestine, the blood first passes to the liver. Here a small amount of alcohol is removed by the liver and excreted in the urine. As it passes through the liver, some alcohol reacts with oxygen in the blood. This produces carbon dioxide and water and provides your body with energy:

alcohol + oxygen → carbon dioxide + water + energy

Most of the alcohol, however, passes through the liver unchanged and gets to the heart, to the lungs (where small amounts pass into the breath) and to the brain. Having gone round the body, some of it returns to the liver and the cycle begins again.

beer

½ pint
284 cm³

alcohol
4%

wine

125 cm³

alcohol
10%

sherry

50 cm³

alcohol
20%

spirits
(whisky
gin
vodka)

25 cm³

alcohol
40%

Figure 4

How does alcohol affect your body?

In general, the more you weigh, the less you are affected by alcohol. This is because the alcohol can spread throughout a large volume of your body. It is less concentrated and therefore has a reduced effect.

Men and boys are also less affected by alcohol than women and girls of the same weight. This is because males have larger livers than females, so they can remove alcohol from their bodies more quickly.

In small amounts, drink can make some people chatty and funny. Others become aggressive.

Who do you think will be affected more by drinking alcohol, a stocky middle-aged man or a slim young woman? To answer this question, you have to consider four main factors:

♦ how much they have drunk,
♦ whether they have had a meal with their drink,
♦ their weight,
♦ their sex.

Alcohol affects your brain and your liver more than other parts of your body.

a) **Alcohol affects your brain**
    Alcohol is a depressant. This means that it depresses (slows down) your thinking. So, it slows down your judgement, your self-control and your skills. You will react more slowly to danger and operate machinery and cars with less care. In the UK, one in every three drivers killed in road accidents are over the legal limit. Excessive drinking can also cause brain damage and psychiatric (mental) problems like depression.

b) **Alcohol affects the liver**
    The liver is like a car with only one gear. It works best at one steady rate. Too much alcohol over a number of years can lead to hepatitis (inflammation of the liver) and cirrhosis (scarring of the liver).

4   Explain why the concentration of alcohol in someone's body is affected by:
♦ how much they have drunk,
♦ whether they have had a meal with their drink,
♦ their weight,
♦ their sex.

Tea is Britain's national drink. On average we drink 3.7 cups of tea a day. Figure 6 below shows the popularity of different drinks and the average number of cups we drink per day.

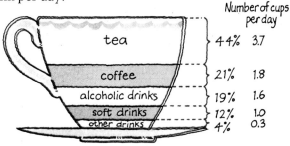

Figure 6   *The most popular drinks*

Table 1

Year	Cost of 1 kg of tea	Average weekly earnings
1700	£1.00 to £4.00	12.5p
1800	£1.80 to £5.50	25p
1900	33p	50p
Today	£3.50	£150

1 Suggest reasons why tea was much cheaper in 1900 than in 1800.

2 Would you think that more people can afford to buy tea today than in 1900?

3 At what date(s) was tea priced as a luxury? Give your reasons.

How is tea produced?

Figure 7

1 Plucking   Tea leaves are plucked from small tea bushes (figure 7).

Figure 8

2 Withering   Plucked leaves are spread on racks in a current of *warm* dry air and left to 'wither' (figure 8). This allows the moisture to evaporate from the leaves.

Figure 10

4 Fermenting   The crushed leaves and juices ferment (figure 10). The green pigment (chlorophyll) in the leaves reacts with oxygen and the leaves turn copper colour.

Figure 9

3 Rolling   The withered leaves are crushed by rolling, cutting and tearing (figure 9). This releases juices from the leaves.

5 Drying   The fermented leaves are dried in *hot* air. The leaves turn black. The leaves are then **sorted**, **tasted** and **packed**.

1 Why is *warm* air used in the 'withering' process?

2 Why are the withered leaves crushed and cut in the 'rolling' process before fermentation?

3 Do you think uncut and uncrushed leaves would ferment? Explain your answer.

4 Why are the fermented leaves dried in *hot* air?

1 All drinks contain water.
◆ Some drinks have water in them already.

◆ Other drinks are made by adding water.

2 We all need water to live.
Communities need large quantities of water for drinking and other uses.
In some countries, water shortage is a severe problem.

3 Drinking water (tap water) usually contains dissolved substances.

4 Drinks such as tea, lemonade and wine, also contain dissolved substances. In these drinks
◆ water is the **solvent**,
◆ the dissolved substances, like sugar in tea, are called **solutes**.

Dissolved substances are described as **soluble**. Substances which do not dissolve are described as **insoluble**.

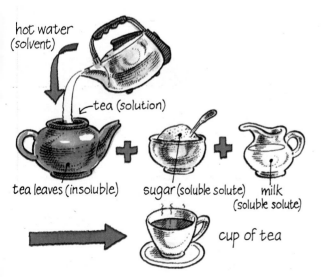

Figure 11 *Making a cup of tea*

5 Water can be made safe to drink by
◆ filtering out insoluble impurities like sand and mud,
◆ killing bacteria by boiling or by adding chlorine.

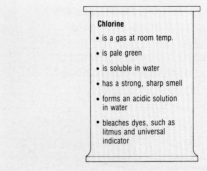

Chlorine
• is a gas at room temp.
• is pale green
• is soluble in water
• has a strong, sharp smell
• forms an acidic solution in water
• bleaches dyes, such as litmus and universal indicator

Figure 12 *Properties of chlorine – a soluble gas used to purify water*

6 Carbon dioxide is added to some drinks to make them fizzy.

Carbon dioxide
• is a gas at room temp.
• is colourless
• is slightly soluble in water
• has no smell
• forms a weakly acidic solution in water
• forms a milky precipitate with lime water
• puts out a burning splint

Figure 13 *Properties of carbon dioxide*

7 Alcoholic drinks are made by fermenting sugary solutions.

8 Alcoholic drinks can be made stronger by **distillation**.

distillation = evaporation then condensation

Different liquids boil at different temperatures. For example, alcohol boils at 78°C and water boils at 100°C. When a mixture containing water and alcohol is heated, alcohol evaporates more easily. When the vapour condenses, the liquid contains a higher percentage of alcohol than the original mixture. This is how spirits like whisky and gin are produced.

9 Although alcoholic drinks are enjoyed by many people, they can result in very serious problems.

10 All substances and materials are made up of **particles**. The preparation and properties of drinks can be explained using the idea of particles (particulate theory).

1 Where does our water come from?

Every time you turn on the tap you expect to get as much water as you need. But where does the water come from? The water from your taps comes via underground pipes from water treatment plants. In England and Wales, 16 200 million litres (3360 million gallons) of water are supplied every day. Where does this treated water come from originally?

Figure 15 *One of the large reservoirs which supply water to people living in the Thames Valley.*

Figure 14 *The water cycle*

Figure 14 shows where water is found on the earth and what happens to it. You can see that it is constantly changing from one form to another in a cycle. Heat from the sun causes surface water to evaporate from rivers, lakes and oceans. This water vapour collects as clouds. As the clouds rise, they cool down. This causes the water vapour to condense as drops of water (or snow flakes).

When it rains, most of the water soaks into the earth as ground water. The rest runs off the land and into streams, rivers and lakes as surface water, much of which seeps back into the oceans.

We take our water from two sources:
a) **surface water** in rivers, lakes and reservoirs (figure 15),
b) **ground water** in underground wells.

Even when surface soil is dry and dusty, porous layers below ground can act like sponges and store vast amounts of water.

2 How is drinking water purified?

Most of the water in our streams and rivers is unsuitable for drinking. It has probably fallen as rain through polluted air, then run along muddy fields or dirty streets. In earlier centuries clean river water was hard to come by and therefore expensive (figures 16 and 17). Nowadays water is cleaned before we use it. This takes place at the waterworks or water treatment plant. Figure 18 opposite is a diagram of a typical water treatment plant. In figure 18 each process is named and its effect is explained.

In some areas, the water is treated further. This often involves the following two processes:
pH adjustment Some waters are acidic enough to react with metal pipes. Lime can be added to neutralise the acid and adjust the pH.
Fluoridation In some areas, about one gram of fluoride is added to every million grams of water (i.e. 1 part per million; 1 ppm). This helps to prevent tooth decay and to reduce bone weaknesses in elderly people.

Figure 16 *The Romans built great aqueducts to carry clean drinking water from springs and rivers into towns and cities. The aqueduct in this photograph was built at Pont du Gard in Southern France.*

Figure 17 *17th century seller of clean river water*

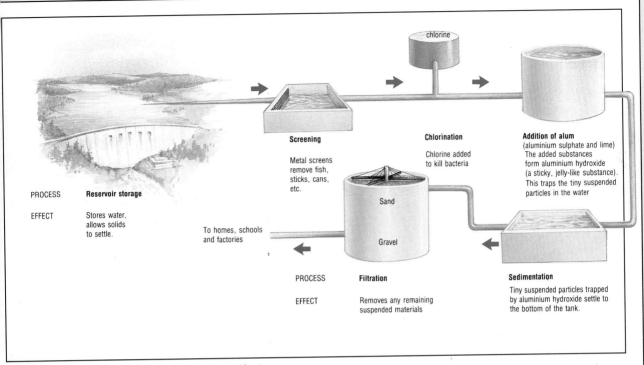

Figure 18 *Processes at a typical water treatment plant*

3 How are alcoholic drinks made?

All alcoholic drinks are made by **fermenting** sugary solutions. People have been fermenting the sugars in honey and fruit juices for at least ten thousand years.

Sugars are carbohydrates. They contain the elements carbon, hydrogen and oxygen. The correct name for table sugar is **sucrose**. This can be used to make alcoholic drinks. Other sugars are also used. These include **maltose** from barley which is used to make whisky. Starch can also be used for alcoholic drinks in place of sugar. This is because it breaks down fairly easily to from maltose. Table 1 shows the source of sugar for different alcoholic drinks.

Table 1 *The source of sugar for different alcoholic drinks*

Drink	Source of sugar	Sugar fermented
Beer	Barley	Starch → maltose
Cider	Apples	Sucrose, maltose and other sugars
Wine	Grapes	Sucrose, maltose and other sugars
Sherry	Grapes	Sucrose, maltose and other sugars
Whisky	Barley	Starch → maltose
Gin	Wheat (corn)	Starch → maltose
Vodka	Potato	Starch → maltose
Saki	Rice	Starch → maltose

Figure 19 *Vat of fermenting lager*

Fermentation needs yeast as well as sugar. Yeast is a micro-organism. The yeast lives on the sugar and splits it up into carbon dioxide and alcohol.

$$\text{sugar} \xrightarrow{\text{yeast}} \text{alcohol} + \text{carbon}$$
$$\text{(sucrose; maltose)} \qquad\qquad\qquad \text{dioxide}$$

Figure 20 shows a simple experiment which you may have used to make a solution of alcohol by fermentation.

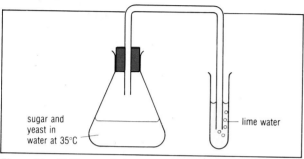

sugar and yeast in water at 35°C

lime water

Figure 20 *Making a solution of alcohol*

After about 30 minutes, the lime water turns milky as carbon dioxide is produced. At the same time, alcohol is left in the flask. However, alcohol is a poison for the yeast. When the alcohol concentration gets to about 15 per cent by volume, all the yeast is killed. Because of this, it is impossible to make alcoholic drinks containing more than 15 per cent alcohol by fermentation alone.

There are two ways of making drinks with more than 15 per cent alcohol.

a) **Distilling** fermented liquids. This method is used to make whisky, gin, vodka and saki. These alcoholic drinks are called **spirits** (table 2).

b) Adding extra alcohol. This method is used to make sherry and port. These drinks are called **fortified wines** (table 2).

Table 2 *The percentage of alcohol in different drinks*

Drink	% alcohol	How it is made
Beer	4	fermented barley
Wine	10	fermented sugars in grape juice
Sherry	20	wine with added alcohol
Whisky	40	distilled liquid from fermentation of barley
Gin	40	distilled liquid from fermentation of wheat (corn)
Brandy	40	distilled wine

4 What happens in terms of particles when tea is made?

i) Boiling the water

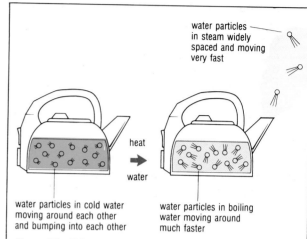

water particles in steam widely spaced and moving very fast

heat

water

water particles in cold water moving around each other and bumping into each other

water particles in boiling water moving around much faster

Figure 21 *What happens to particles when water is boiled?*

ii) Brewing the tea

hot water poured on to tea leaves

tea is brewed

hot water particles moving about rapidly and bumping into the tea leaves

some particles in tea leaves are 'knocked off' the surface of the tea leaves and dissolve in the water

10 minutes

tea leaf

particles of substances in tea leaves vibrating (jittering) about fixed points

some particles in tea leaves are insoluble in the water and stay as solid

Figure 22 *What happens to particles when tea is brewed?*

iii) Adding milk to the tea

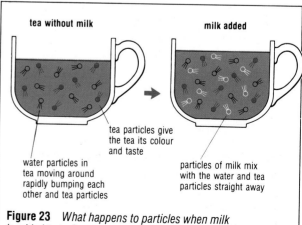

tea without milk

milk added

tea particles give the tea its colour and taste

water particles in tea moving around rapidly bumping each other and tea particles

particles of milk mix with the water and tea particles straight away

Figure 23 *What happens to particles when milk is added to tea?*

Things to try out

1 *Do a survey.*
   a) Carry out a survey of drinks among your friends and your family. Ask them what drinks they have had in the last 24 hours. Make a bar chart or a pie chart of your results.
   b) Carry out a survey of the popularity of different fruit juices among your friends and your family. Your survey could include the following juices: apple, grapefruit, orange, pineapple, tomato. Make a bar chart or a pie chart of your results.

2 *Make a drink from oranges or lemons.*
   a) Plan a recipe for an orange or lemon drink.
   b) Now compare your recipe with one in a cookery book.
   c) What changes will you make to your recipe?
   d) Make the drink.

3 *Collect cuttings related to drinks.*
   a) Collect cuttings related to drinks from magazines or newspapers.
   b) Write six sentences which pick out important points in the cuttings.

4 *Design a poster.*
   Design a poster which could be used as an advert for your favourite drink.

5 *Make tea.*
   a) Plan an experiment to see whether there is any difference between tea made
      A. by putting milk into a cup first and then pouring in the tea or
      B. pouring out the tea and then adding the milk.
      Describe
        i) the experiment itself,
        ii) the care you will take to make the comparisons fair,
        iii) the tests you will use to check whether there is any difference.
   b) Carry out your experiment.
   c) What are your conclusions?

Things to find out

6 Find out about the water supply to your home. Some questions to consider are given below.
   a) How much do you pay in water rates?
   b) Where does your water come from?
   c) Where is the nearest water treatment plant to your home?
   d) What other services does your water company provide besides simply supplying water?
   e) How does your water supply company spend the money you pay in water rates?

7 Use a library or an encyclopaedia to find out about one of the following drinks:
   milk, coffee, fruit juice, beer, whisky.
   Some possible questions to consider are:
   a) How is the drink made or processed before it goes on sale?
   b) What chemicals (constituents) does the drink contain?
   c) What regulations cover the labelling and sale of the drink?
   d) How have sales of the drink changed during the last ten years or so?
   e) What arrangements are there for returning or recycling the containers for the drink?

8 Find out the solute(s) and solvent in the following solutions:
   a) brine   b) vinegar
   c) wine   d) lemonade
   e) tap water   f) liquid paper

Points to discuss

9 Suppose that two of your best friends have got into the habit of drinking heavily. What advice would you like to give them?

10 Some people think it is wrong to have any drinks which contain alcohol. What do you think?

11 Helen who is fifteen years old says, 'It's just being sociable to go to the pub for a drink. If I were the only one having a coke, I'd feel left out.' What do you think?

12 Some people believe that it is more important to improve the water supply and water treatment in poorer countries than to improve the hospital care and education in these countries. What do you think?

Questions to answer

13 Describe the experiments that you would carry out in order to:
   a) find the amount of insoluble solid suspended in 10 cm^3 of a sample of muddy water,
   b) find the amount of solids dissolved in 10 cm^3 of clean river water.

14 Micro-organisms such as yeast can obtain energy from foods such as sugar. If there is no oxygen present, the yeast lives on the sugar producing carbon dioxide and alcohol. This process is called **fermentation**. Two students set up the apparatus in figure 24 at room temperature. They were trying to study the best temperature for fermentation.

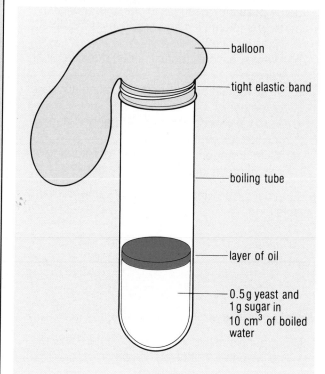

balloon

tight elastic band

boiling tube

layer of oil

0.5g yeast and 1g sugar in 10 cm³ of boiled water

Figure 24

a) What is the balloon for?
b) Why is the tight elastic band used?
c) What happens to the balloon during the experiment?
d) Why did the students use boiled water?
e) Why is the fermenting mixture covered with a layer of oil?
f) Write a word equation for the fermentation process.
g) What should the students do now to find the best temperature for fermentation?

15 Explain the following observations using the idea of particles.
a) Sugar dissolves in water to form a clear solution.
b) Tea has its own special smell.
c) Sugar dissolves more quickly in hot tea than in cold tea.

16 Sugar solutions are often used in cooking. When sugar dissolves in water to form a solution, sugar is the solute and water is the solvent. Figure 25 shows how the percentage of sugar in the solution affects its boiling point.

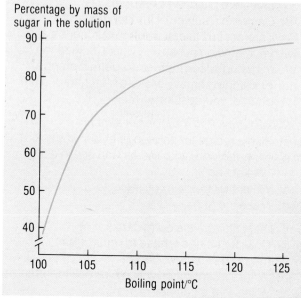

Percentage by mass of sugar in the solution

Boiling point/°C

Figure 25

a) What happens to the boiling point of the sugar solution as the percentage of sugar increases?
b) Why do you think the percentage of sugar affects the boiling point?
c) Sugar solutions are boiled to make jam and fondant icing. In jam making, fruit, sugar and water are boiled to 105°C. Then the jam is allowed to set. When fondant icing is made the sugar solution is boiled at 114°C before cooling. What is the concentration of sugar in i) jam, ii) fondant icing?

17 Read section 2, on page 18. *How is drinking water purified*?
a) Make a summary of the stages in purifying our water supplies.
b) In many areas, water is taken from a river *above* a town and used water (**effluent**) is put in the river *below* the town. Why is this?
c) Why do some dentists tell their patients to use fluoride toothpaste?

18 Suppose you are a hospital doctor. One of your patients seems to be urinating a much higher proportion of the liquid that he drinks than is normal. How would you check this?

Introducing warmth

No-one enjoys being cold. In fact being cold can be fatal. Every year we hear of old people who die of cold in the winter. How do we keep ourselves warm?

◆ We eat food which gives us energy.
◆ We keep our homes, schools and workplaces warm.
◆ We try to avoid losing heat unnecessarily.
◆ We insulate our bodies with clothes and insulate the places where we live.

Figure 1   *In very cold climates special clothing must be worn. How does this protect the body?* ▶

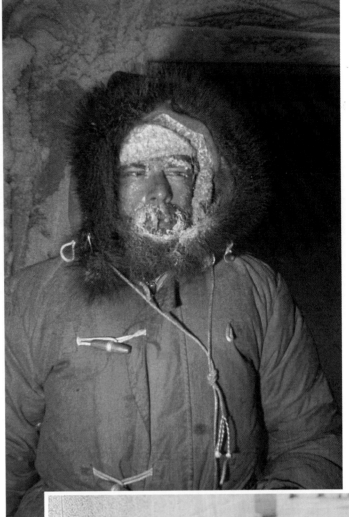

▲ **Figure 2**   *What can old people do to keep warm?*

Figure 3   *How do you heat your home? Think of as many ways of heating a home as you can.* ▼

Figure 4   *Marathon runners are sometimes wrapped in aluminium covers after the race to help them conserve energy.* ▶

In this chapter you will see how

◆ chemistry can help us provide warmth more conveniently and cheaply,
◆ chemical changes take place when fuels burn,
◆ our environment is affected when we burn fuels.

1 Heating our homes

We are lucky in Britain because we have plentiful energy supplies. We have **coal**, and **oil** and **gas** from the North Sea. We can use these energy sources to make electricity.

Different people choose different energy sources to warm their homes.

Figure 5 *Coal or wood is burned in open fires to provide heating and hot water in the home.*

Figure 6 *This looks like a coal fire but the fuel is gas – a coal-effect fire.*

Figure 7 *Electricity is used in electric heaters. Storage heaters use cheaper off-peak electricity.*

Figure 8 *The oil being drilled in the North Sea is burned in power stations to produce electricity. It can also be burned in domestic boilers to provide cental heating.*

Energy source	Cost per megajoule/pence	Advantages	Disadvantages
Coal	0.40	Cheap Coal fires look nice	Dirty Difficult to transport Makes a lot of smoke and ash Difficult to light
Oil	0.48	Easy to transport and store Can be pumped automatically	Needs to be delivered every so often Messy and smelly if it leaks Price is liable to vary
Gas	0.35	Doesn't need to be delivered Can be pumped automatically Cheap Clean	Dangerous if it leaks Supply has to be laid to the house
Electricity	1.38 (0.57 off-peak)	Easily switched on and off Very clean Doesn't need to be delivered	Expensive

Table 1 *Information about different energy sources. The costs of the fuels are in pence per megajoule. A megajoule is the amount of energy given out by a small electric fire in about fifteen minutes*

Use table 1 to decide the best energy source for each of the people in examples 1 – 3 to heat their homes.

1 Mrs Evans, aged 72, who lives alone in an isolated cottage in Wales. She has only her pension to live on, and she isn't in the best of health.

2 The Patel family, who live in Manchester. The family is four in number, with a good income.

3 The Waddington family, who live in an enormous, draughty house in Yorkshire. They have only a tiny income.

4 Why is electricity so much more expensive than the other energy sources?

2 Designing a gas burner

Many people use natural gas to heat their homes and to do their cooking. You use natural gas as a fuel every time you use a Bunsen burner.

Natural gas is mainly **methane**. Methane contains the elements carbon and hydrogen. When methane burns in plenty of air, carbon dioxide and water are formed.

methane + oxygen → carbon dioxide + water
$$CH_4(g) + 2O_2(g) \rightarrow CO_2(g) + 2H_2O(l)$$

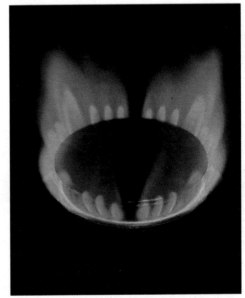

▲ **Figure 9** *The flame on a gas cooker is like a Bunsen flame.*

◄ **Figure 10** *What kind of Bunsen flame is shown here?*

25

If there is a shortage of air, there is not enough oxygen to form carbon dioxide. Poisonous carbon monoxide (CO) may be formed instead. If the air supply is really bad, the carbon may not oxidise at all. It may just form soot. This soot will be deposited on whatever you are heating.

What does an efficient gas burner need?

An efficient gas burner needs two features.

◆ A controlled flow of gas which can be turned on and off and adjusted.
◆ A good supply of air.

Figure 11a shows how a Bunsen burner gets its air supply when the air hole is open. The flame is said to be **aerated**. Figure 11b shows how it gets it with the air hole closed. The flame is **non-aerated**.

1 What colour is the flame when the air hole of a Bunsen burner is
   a) open, b) closed?
2 When you are heating things with a Bunsen burner, you should do it with the air hole open. Think of *two* disadvantages of heating things with the air hole closed.
3 Think about the flame on a gas cooker. Is it aerated or non-aerated? How do you know?

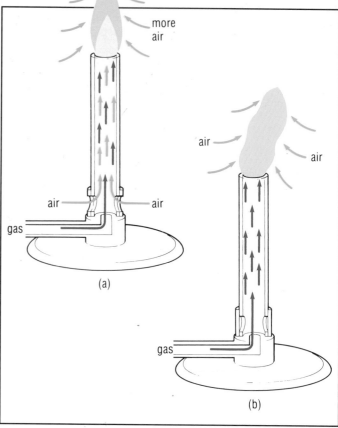

Figure 11
a) Air hole open. Aerated flame: gas already mixed with air when it burns.
b) Air hole closed. Non-aerated flame: gas gets air supply from air round flame.

The design of gas cooker burners

A gas cooker burner needs to give a clean flame so that it doesn't make saucepans sooty on the outside. The flame must be easily controlled. Figure 12 shows the design of a typical burner.

4 In what ways is this burner similar to a Bunsen? In what ways is it different?

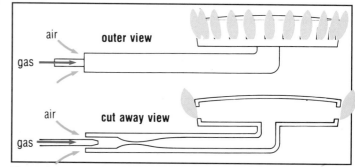

Figure 12   A typical gas burner

Design a burner

The Northern Glass Company makes glass vases. One of the steps in the manufacture of a vase involves heating a thick solid glass rod evenly (figure 13).

Gas is used to heat the rod.
Sketch the design of a burner that would do this job.

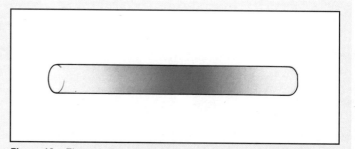

Figure 13   The rod has to be heated evenly all round the shaded region.

3 Fire!

Home fires kill about 800 people each year in Britain. Fires are often easy to light, but difficult to put out.

The fire triangle

A fire needs three things:

◆ fuel – anything that will burn,
◆ oxygen – usually from the air,
◆ a high temperature to start the fire and keep it going.

These three things make up the **Fire Triangle** (figure 14). You can put out a fire by taking away any one of the three.

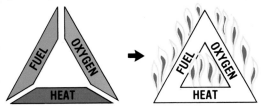

Figure 14   *The Fire Triangle*

Figure 15   *What fire-fighting methods would be used to fight a fire like this?*

How to put out fires

Table 2 gives some of the methods that can be used to put out fires.

Table 2   *Fire-fighting methods*

Fire-fighting method	What it involves	Where you get it
Fire blanket	A blanket made of non-flammable material is thrown over the fire	Red-painted fire blanket container
Sand	Sand is thrown on the fire	Red-painted fire bucket
Water	Cold water is poured over the fire. Cannot be used on fires involving oil, petrol or electricity	Tap, fire hose or red-painted water-type fire extinguisher
Carbon dioxide	Carbon dioxide gas is directed at the fire, and forms an invisible blanket over it	Black-painted carbon dioxide-type fire extinguishers
Foam	A foam containing bubbles of carbon dioxide is poured over the fire	Cream-painted foam-type fire extinguishers
Halon	A non-flammable liquid made from carbon, chlorine, bromine and fluorine is poured on the fire. It forms a dense blanket of vapour	Green-painted halon-type fire extinguishers
Powder	A fine powder is poured over the fire	Blue-painted powder-type fire extinguishers

Use table 2 to answer these questions.

1 Each of the fire-fighting methods in the table works by taking away one or more of the three sides of the Fire Triangle. For each method, decide how it works.

2 Why must water not be used against oil, petrol or electrical fires?

3 Look around your chemistry laboratory at school. List the different kinds of fire-fighting equipment in the laboratory.

4 What would you do in each of the following situations?
  a) You are having a bonfire in the garden. The fire gets out of control and the hedge starts to burn.
  b) You are cooking chips when the chip pan catches fire.
  c) You are babysitting when a small girl's nightdress catches fire.
  d) You are in the laboratory studying the Warmth chapter. You are testing the value for money of methylated spirit as a fuel. Suddenly the boy next to you spills a bottle of methylated spirit. It spreads over the bench and catches light.

4   Cleaning up the power stations

One of the worst **air pollutants** is **sulphur dioxide**, (SO_2). Chemists believe that sulphur dioxide is one of the major causes of **acid rain**. Acid rain can easily damage plants and animal life as well as buildings. Coal-fired power stations are major sources of sulphur dioxide poisoning. These power stations burn coal to generate electricity. Coal contains about 1.5 per cent sulphur. The sulphur is in the form of compounds. When coal burns, the sulphur burns too, forming sulphur dioxide. Nearly half of the sulphur dioxide given out by human activities each year comes from power stations.

The people who run the power stations are trying to find ways of cutting down the amount of sulphur dioxide given out. They have looked at several possible solutions. These include:

◆ Using coal with less sulphur in it. This would have to come from abroad.
◆ Building nuclear power stations to replace coal-fired stations.
◆ Getting the sulphur out of the coal *before* burning it.
◆ Getting the sulphur dioxide out of the gases *after* burning the coal.

In fact they chose the last solution – getting the sulphur dioxide out of the gases after burning the coal. This is called **flue gas desulphurisation** (FGD) (figure 17).

How does FGD work?

Sulphur dioxide is an acidic gas. Like all acids, it is neutralised by alkalis. The alkali used is lime – calcium oxide (CaO). This is a very cheap alkali which is made from limestone.

The acidic sulphur dioxide reacts with the alkaline lime. This forms calcium sulphite ($CaSO_3$), which is a solid.

sulphur	+	calcium oxide	→	calcium
dioxide		(lime)		sulphite
$SO_2(g)$	+	$CaO(s)$	→	$CaSO_3(s)$

The calcium sulphite is then reacted with air. If forms calcium sulphate ($CaSO_4$), which is also called **gypsum**.

calcium	+	oxygen	→	calcium
sulphite		(air)		sulphate
$2CaSO_3(s)$	+	$O_2(g)$	→	$2CaSO_4(s)$

Figure 16   *Fitting a flue gas desulphurisation plant is expensive. For a big power station like this it would cost about £200 million to fit. It would also cost about £30 million a year to run.*

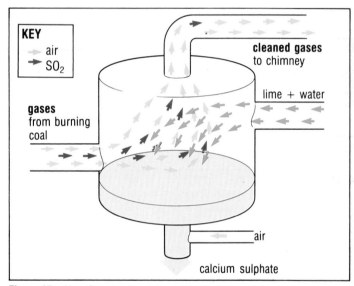

Figure 17   *How flue gas desulphurisation (FGD) works*

Calcium sulphate is a useful chemical. It is used to make plaster. Plaster is important in the building trade.

1   Suppose all Britain's coal-fired power stations were fitted with FGD. They would produce about 8 million tonnes of calcium sulphate a year. The building industry needs about 3 million tonnes of calcium sulphate a year. What would happen to the rest?

2   Where would the money for fitting FGD come from? Who would pay for it? Do you think it is worth the expense?

1 Fuels burn in air, giving out heat. This is called combustion. A reaction which gives out heat is called an exothermic reaction.

2 Combustion needs three things: fuel, oxygen and heat. The rate of combustion depends on the conditions, particularly the concentration of oxygen. Fuels burn much faster in pure oxygen. Sometimes they burn so fast that they explode. (There is more on combustion in the *Thinking About* section.)

3 Combustion is a chemical reaction. A chemical reaction always involves the formation of a new substance. Fuel reacts with oxygen to form new substances. Most fuels contain carbon and hydrogen, and these combine with oxygen to form carbon dioxide and water when the fuel burns. This process is called oxidation. Fuels may also form other, unwanted substances when they burn (figure 18).

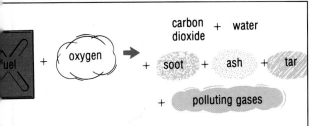

Figure 18 *Products of burning fuel*

4 Different fuels have different properties. Figure 19 shows some of them.

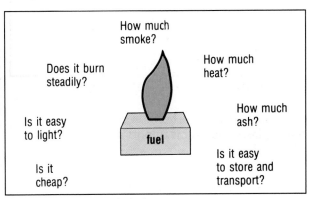

Figure 19 *Properties of a fuel*

5 Coal burns to form soot, ash, tar and acidic gases.

6 Different fuels are suitable for different purposes. We choose a fuel for a particular job according to its properties. Cost, convenience and pollution are particularly important when deciding which fuel to use. No fuel is completely ideal.

7 Fuels are often processed to improve them. Coal is processed to make smokeless fuels such as **coke**. Crude oil is processed to give **petrol**, **paraffin** and other fuels.

8 Pollution is the contamination of the environment by substances made by humans. Combustion of fuels often causes air pollution.

9 Figure 20 shows some of the pollutants that may be formed when fuels burn. The amounts of pollutants that are formed depend on the fuel, and the way it is burned.

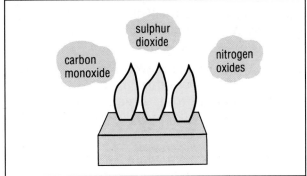

Figure 20 *Some common pollutants formed when fuels burn*

10 Many fuels, particularly coal and coke, contain sulphur. When they burn, the acidic gas sulphur dioxide is formed. This can help form **acid rain**, as figure 21 shows.

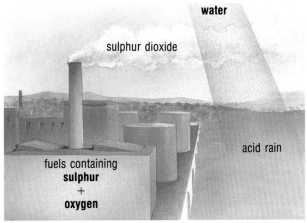

Figure 21 *Sulphur-containing fuels can cause acid rain*

11 Acid rain is harmful to living things. It is harmful to trees, and to life in rivers and lakes. Acid rain and sulphur dioxide in the air cause damage to buildings. They make metals corrode faster.

12 Pollution can be cut down, but this costs money. We need to balance the cost against the benefits of controlling pollution.

1 What is energy?

We depend on energy to keep things going. Without energy, life would grind to a halt.

Energy comes in different forms. Figure 22 lists some important forms of energy.

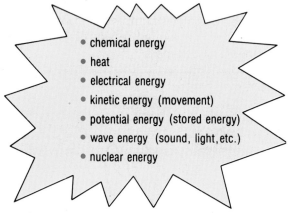

- chemical energy
- heat
- electrical energy
- kinetic energy (movement)
- potential energy (stored energy)
- wave energy (sound, light, etc.)
- nuclear energy

Figure 22 *Important forms of energy*

Energy can be converted from one form to another. Figure 23 shows the energy conversions which go on in a coal-fired power station.

coal burns → steam drives turbines → turbines drive generators

| CHEMICAL ENERGY | → | HEAT | → | KINETIC ENERGY | → | ELECTRICAL ENERGY |

Figure 23 *Energy conversions in a power station*

Where do we get energy from?
Society needs energy sources to keep things going. The most important energy sources are **fuels**. Fuels contain stored chemical energy. They burn in air, releasing the chemical energy as heat. This heat can be used to keep us warm, or to drive motor vehicles. It can drive power stations which make electricity.

Food is a fuel. We need it to provide the energy to keep our bodies going.

2 What happens when fuels burn?

Another name for burning is **combustion**. During combustion, a fuel reacts with oxygen. This oxygen usually comes from the air.

Most fuels contain the elements carbon and hydrogen. When the fuel burns, the carbon and hydrogen are oxidised. They form carbon dioxide and water. For example, with petrol:

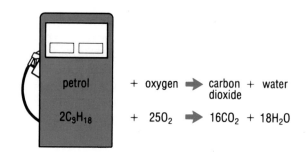

petrol + oxygen $\rightarrow$ carbon dioxide + water

$2C_9H_{18}$ + $25O_2$ $\rightarrow$ $16CO_2$ + $18H_2O$

Air is only about one-fifth oxygen. The rest is mostly unreactive nitrogen. Only the oxygen is involved in combustion. The nitrogen does not react.

Fuels burn much more fiercely in pure oxygen than they do in air. Figure 24 shows an oxy-acetylene burner. Acetylene is a gas which burns in air, rather like natural gas. But in pure oxygen it burns much more fiercely. An oxy-acetylene flame is hot enough to cut through steel.

Figure 24 *Oxy-acetylene torches are used for cutting through steel*

What happens when there isn't enough oxygen?

Sometimes fuels cannot get all the oxygen they need to burn. For example, in a car engine the petrol may not get enough air to burn properly.

When the oxygen supply is poor, fuels burn to give different products. With less oxygen available, the carbon in the fuel forms carbon monoxide instead of carbon dioxide. Carbon *monoxide* contains only half as much oxygen as carbon *dioxide*.

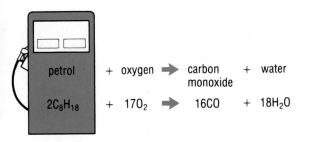

petrol + oxygen ➡ carbon + water
monoxide

$2C_8H_{18}$ + $17O_2$ ➡ $16CO$ + $18H_2O$

If the oxygen supply is really bad, there may not even be enough to form carbon monoxide. The carbon stays unchanged and unoxidised. It comes off as black, sooty smoke.

Carbon monoxide is a dangerous gas, because it is very poisonous. It stops the blood carrying oxygen properly. Your body can cope with small amounts of carbon monoxide, but large amounts can kill.

Cigarette smoke contains a lot of carbon monoxide (figure 25). This is one of the many reasons why smoking is so bad for your health.

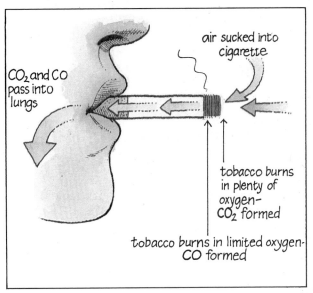

air sucked into cigarette

CO_2 and CO pass into lungs

tobacco burns in plenty of oxygen– CO_2 formed

tobacco burns in limited oxygen– CO formed

Figure 25 *Why cigarette smoke contains carbon monoxide*

Carbon monoxide is also present in car exhaust gases. It may also be given off by gas burners and paraffin heaters which have their air supply blocked with dirt.

Taking it further

To survive, your body must have a constant supply of oxygen. Oxygen is carried around the body in the blood. Blood contains a substance called **haemoglobin**. Haemoglobin combines with oxygen, but the bond between them is weak.

When blood passes into the lungs, lots of haemoglobin molecules form weak bonds to oxygen molecules, forming **oxyhaemoglobin**. When the blood arrives in other parts of the body, where oxygen is in short supply, the oxyhaemoglobin breaks down. Haemoglobin is reformed, and oxygen is released. The haemoglobin is now free to pick up another oxygen molecule in the lungs (figure 26a).

The bad news is that carbon monoxide can also combine with haemoglobin, to form **carboxyhaemoglobin**. And the bond between carbon monoxide and haemoglobin is *strong*. So carbon monoxide cannot be easily removed from the haemoglobin. This stops the haemoglobin being

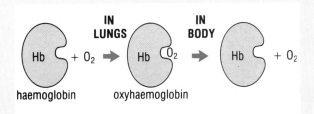

IN LUNGS **IN BODY**

Hb + O_2 ➡ Hb O_2 ➡ Hb + O_2

haemoglobin   oxyhaemoglobin

Figure 26 *a) Haemoglobin and oxygen*

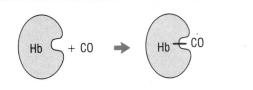

Hb + CO ➡ Hb–CO

Figure 26 *b) Haemoglobin and carbon monoxide*

able to carry oxygen. If too many haemoglobin molecules get blocked by carbon monoxide, the blood cannot carry enough oxygen to keep the body going.

4 Why do burning fuels cause pollution?

Fuels produce gases when they burn. The main gases are

◆ Carbon dioxide (CO_2) formed when carbon in the fuel is oxidised.
◆ Sulphur dioxide (SO_2) formed when sulphur impurities in the fuel are oxidised.
◆ Nitrogen oxides (NO and NO_2) formed when nitrogen and oxygen in the air combine together.

All chemical elements can be classed as metals or non-metals. Metals have very different properties from non-metals. Their oxides have very different properties too. In particular

◆ Metal oxides are always solids.
  Non-metal oxides are often gases or liquids.
◆ Metal oxides are alkaline.
  Non-metal oxides are acidic.

You can see why non-metal oxides like SO_2 and NO_2 cause such problems when fuels are burned. Not only are these oxides acidic. They are also gases, so they escape into the air and cause pollution.

Why do acid gases cause damage?
Acid gases are most damaging when they combine with rain water. They react, forming acidic solutions.

◆ Sulphur dioxide forms sulphurous acid (H_2SO_3) and sulphuric acid (H_2SO_4).
◆ Nitrogen oxides form nitric acid (HNO_3).

These acidic solutions are only very dilute, but they are still very corrosive. They attack stone and metal, and are harmful to plant and animal life.

Figure 27 *Non-metal oxides escape into the air from the chimney of a coal-fired power station*

Taking it further

Figure 28 shows the types of oxides formed by elements in different parts of the Periodic Table.

1 Whereabouts in the Periodic Table are metals to be found?
2 Whereabouts are non-metals to be found?
3 What kind of elements are found in the middle of the table?

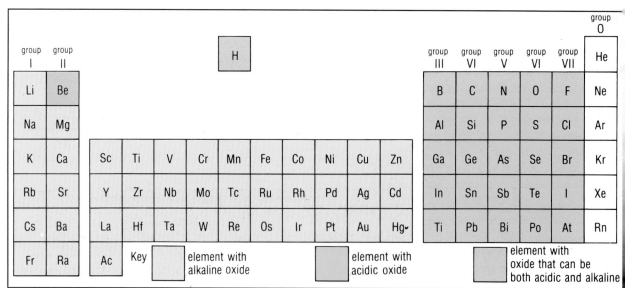

Figure 28 *Types of oxides formed by elements in different parts of the Periodic Table*

Things to try out

1 *Were they the good old days?*
For this activity you will need to talk to a person who is over 70 years old.

Ask the person what things were like when they were your age. Ask them
a) What fuel was used to warm their home?
b) Was every room heated? If not, which rooms were heated?
c) Which fuel was used for cooking food?
d) What means of transport did they use:
i) for journeys less than 2 miles or so,
ii) for journeys over 2 miles or so?

2 *Making a camp fire*
You're out on a picnic, but you've forgotten the camping stove. It is a windy day. You have to make a fire to boil the kettle. You can use only the materials shown in figure 29. You don't necessarily have to use all of them.

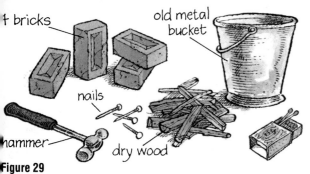

Figure 29

Sketch your camp fire design. Briefly explain why you chose this design.

Things to find out

3 What is self-warming food? How does it work?

4 The fuels listed below are not very common in Britain. For as many as you can, find out what they contain, and where they are used.
a) Charcoal b) Peat c) 'Meta' fuel
d) Bagasse

5 Petrol, paraffin, diesel oil and fuel oil are all made from crude oil. How?

6 Some people use solar panels. How do they work?

7 North Sea Gas has only been in use since the 1970s. What gaseous fuel was used before then? Where did it come from?

Making decisions

8 *Energy sources for industry*
Look back at table 1 on page 25. Which energy source would be best for each of the following? Give a reason for each choice.
a) Fuelling a power station in Yorkshire.
b) Fuelling a power station in Saudi Arabia.
c) Heating an oven in a large bakery.
d) Heating steel bars in a steelworks, before rolling them to make steel plate.
e) Heating a large greenhouse used for growing tomatoes.

Points to discuss

9 It's a winter evening and you're sitting watching television. You're wearing a T-shirt and jeans. The heating is on, but even so you begin to feel a bit chilly. Do you
a) put on a pullover, or
b) turn up the heating?
Which is the most sensible choice? Why?

10 In winter, many old people in Britain suffer from the cold. They may even develop a condition called hypothermia. They get so cold their bodies stop working properly.
Why do you think old people are particularly likely to suffer from the cold?

11 Suppose you are Minister for the Environment. What laws would you pass to try and cut down air pollution? Remember – laws have to be *enforceable*. You have to be able to prove they are being broken. Remember too that it is impossible to remove air pollution *completely*.

12 Controlling acid rain could put up the price of electricity in Britain. Why? Do you think British people would be prepared to pay?

13 CHEMCO is a company which makes chemicals in a small town in the Midlands. CHEMCO employs 400 local people.
CHEMCO is in trouble because the factory is giving out polluting gases. Local residents have complained about the smell. They say pollution is attacking their houses and spoiling their garden crops.
CHEMCO says that to control the pollution would cost £20 million. They could not afford this without laying off one tenth of the workforce.

Discuss what you think the local residents should do.

Questions to answer

14 Copy the following paragraph and fill in the missing words or groups of words. Each word or group of words is used only once. The missing words and groups of words are: **carbon dioxide, carbon monoxide, combustion, exothermic, oxides of nitrogen, sulphur dioxide, water.**

Fuels are substances which burn in air, giving out heat. Another name for burning is ___(a)___. Reactions which give out heat are described as ___(b)___. Most fuels contain the elements carbon and hydrogen. When the fuel burns in plenty of air, the carbon in it forms ___(c)___. But if the air supply is limited, ___(d)___ is formed instead. The hydrogen in the fuel forms ___(e)___ when it burns. Burning fuels can cause air pollution. Common pollutants are ___(f)___ and ___(g)___.

15 Three plastic bottles were filled with different gases, as shown in figure 30.

Figure 30

The three bottles were put behind a safety screen. For each bottle in turn, the stopper was removed and a long burning taper was held over the open mouth.

a) This experiment must be carried out behind a safety screen. Why?
b) What would you expect to happen with each bottle?
c) Explain why the gas in each bottle behaved in this way.
d) A gas leak in a house can be very dangerous. Explain why.

16 You have been given a white solid material. You are told it is a fuel, and that it costs about the same as coal.
   a) What properties would you want the solid to have in order to be a good fuel?
   b) What tests would you do to decide whether it had these properties?

17 Sharon did an experiment to compare wood and charcoal as fuels. She wanted to know which fuel caused least air pollution. The apparatus she used is shown in figure 31.

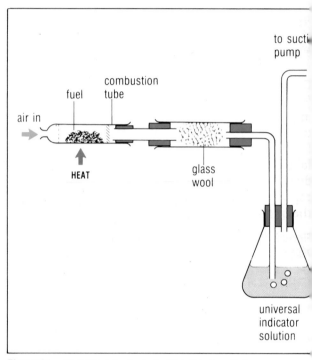

Figure 31 *Apparatus for investigating the air pollution caused by different fuels*

Her results are shown in table 3.

Table 3

Fuel	Appearance of glass wool afterwards	Colour of indicator solution afterwards
Wood	yellow-brown, tarry	yellow (pH 6)
Charcoal	clean	yellow (pH 6)

a) Describe how the apparatus works. Your description should begin 'Air is drawn into the combustion tube ...'
b) What does the experiment tell you about wood as a fuel, compared with charcoal?
c) What other things would you need to know about wood and charcoal before you could decide which was the better fuel?

Introducing clothing

People wear clothes for all sorts of reasons – to keep warm, to keep cool, to keep dry – and to look good! Clothes are usually made up from pieces of fabric, which is another name for cloth. Fabric is built up from thin strands, or fibres, like cotton, wool, nylon and polyester woven together.

Figure 1 *What are the advantages and disadvantages of metal clothing? Apart from knights, who else might wear metal clothing?* ▶

Chemists help to make better clothing for you. They design new fibres, and improve old ones. They find ways to treat fabrics, to make them waterproof for example. They also make dyes to give fabrics attractive colours. What properties do clothing fabrics need? Look at the captions for figures 1 – 4 and think about the questions.

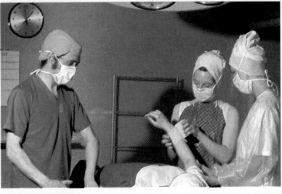

Figure 3
Think of five useful properties of the clothing fabrics you are wearing at the moment. ▼

Figure 2 *What are the advantages and disadvantages of paper clothing?* ▶

Figure 4
What are the advantages and disadvantages of plastic clothing? ▶

In your answers to these questions, you probably thought about the properties of the materials involved. For example, you know that metals feel cold and paper burns.

In this chapter you will see how

◆ chemistry can help us understand the properties of clothing materials,
◆ chemists can use their skills and knowledge to design even better materials.

1 Dyes for brighter clothing

Most fibres are naturally white, or else a rather dull colour. Dyes are used to give bright colours to clothing.

Dyes are coloured substances. But any old coloured substance will not do – the colour must stick to the cloth. For centuries, people used natural dyes. For example, a blue dye was made from the indigo plant. But the colours of natural dyes are often dull and the range of colours is limited. Also, natural dyes are not **fast** – they fade when repeatedly washed or exposed to sunlight.

Nowadays there is a huge range of synthetic dyes, of any colour you can imagine. What's more, modern synthetic dyes are fast.

In 1856, William Perkin was trying to find a way of making the drug quinine, which is used to treat malaria. Instead of quinine he got a beautiful purple dye, which he called **mauve**. Perkin found that mauve dyed cloth, and was fast to light. The dye became famous and very fashionable.

Perkin became rich because of his discovery. Soon many other synthetic dyes of different colours were discovered by chemists. Chemists learnt more about the chemistry of dyes. This made it possible to produce new dyes on purpose instead of by accident. Nowadays, chemists can make practically any colour dye.

Making dyes stick

For a dye to be fast and not wash out, it must stick to the cloth. Like all substances, dyes contain tiny particles. These particles are attracted to the fibres and stick to them, as shown in figure 9.

▲ **Figure 5** *Many dyes have been used to colour these fabrics.*

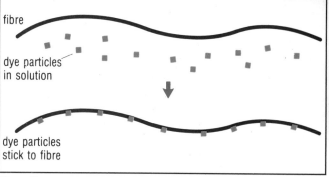

◀ **Figure 6** *Indigo is used to dye blue jeans. But the blue fades with washing and sunlight.*

◀ **Figure 7** *The first synthetic dye was made by an English chemist, William Perkin. He made it by mistake in 1856. He was just 18 years old at the time.*

Figure 8 *Perkin's dye called 'mauve' was celebrated in the Penny mauve stamp of 1881.*▼

fibre

dye particles in solution

dye particles stick to fibre

Figure 9

A particular dye is usually only attracted to a particular type of fibre. This means different fibres need different dyes (figure 10).

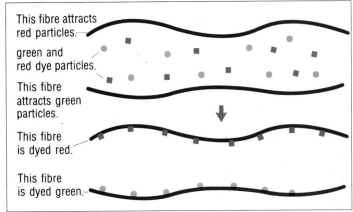

This fibre attracts red particles.

green and red dye particles.

This fibre attracts green particles.

This fibre is dyed red.

This fibre is dyed green.

1 Why do you think some dyes are fast while others run?
2 When you wash a brightly-dyed garment for the first time, it is best to wash it on its own, instead of with other clothes. Why?

Figure 10 *Different dyes stick to different fibres.*

2 Dry-cleaning

Some clothes cannot be washed in water. This is because they would shrink or change shape in water. These clothes are labelled to show how they have to be dry-cleaned (figure 11). Dry-cleaning is also used to remove dirt and stains that water and detergents cannot remove.

Dry-cleaning uses special solvents. These solvents are particularly good at dissolving grease. But they have to be non-flammable – they must not burn, because this could cause a dangerous accident. The solvents must not be poisonous, otherwise they could harm the dry-cleaning workers or the owners of the clothes.

A dry-cleaning machine is really like a big washing machine, but it uses a solvent instead of water. The solvent it uses is recycled and used again and again. It is recycled by distilling. Dirty solvent is boiled, then its vapour is condensed by cooling, to give clean solvent. The dirt is left behind as a solid and can be thrown away.

The fact that the solvent has to be distilled means it needs another property. Its boiling point must not be too high, otherwise it would be difficult and costly to boil. On the other hand, the boiling point cannot be too low, otherwise it would vaporise in the machine.

Chemists can 'design' dry-cleaning solvents so they have exactly the properties that are needed. A solvent that is often used contains the elements carbon and chlorine. It is called tetrachloroethane. It is a colourless liquid, boiling point 121°C.

◄ **Figure 11** *The symbols used on a garment label show how it should be cleaned – it must be ironed with a cool iron, should not be bleached and should be dry cleaned.*

Figure 12 *Dirty residue being removed from a dry cleaning machine. This residue is left after the dirty solvent has been purified.*

1 Petrol is good at dissolving grease. Why is it not used as a dry-cleaning solvent?
2 Is dry-cleaning really 'dry'? Explain your answer.
3 When you take clothes home from the dry-cleaners in your car, you are advised to keep the car windows open. Why?

3 Protective clothes for firemen

Firemen's protective clothes obviously have to be **fireproof**. They also have to be good insulators, to stop heat reaching the body. They have to be waterproof because of all the water that gets pumped onto fires. They have to be durable so they wear well, and also have to be comfortable to wear (figure 13).

It is difficult to get clothing that is fireproof, insulating *and* waterproof. Some plastics *could* be used, but they would melt and stick to the skin at high temperatures.

For many years, firemen wore heavy woollen tunics. As you may know if you have done tests on fibres, wool does not burn easily. Firemen's tunics used to be made of thick, dense wool. The wool was matted so closely that it insulated the body from heat. It was also difficult for water to penetrate. But even so, the wool absorbed a lot of water. Once wet, these jackets became uncomfortably heavy to wear.

So fibre chemists looked for new answers. They found two. One used a natural fibre, one a synthetic fibre.

1   Natural fibre: specially treated wool

The new, lighter weight firemen's tunics are made in three layers as shown in figure 14.

The outer layer is wool, specially treated to make it fireproof and waterproof. The fireproofing treatment is a compound of the element zirconium. The waterproofing treatment uses a fluorocarbon – a non-flammable, water-repellent compound containing carbon and fluorine.

The middle layer is dense, matted wool – as in the traditional jacket, but thinner. It insulates the body against heat. The inner layer is cotton, for comfort next to the skin.

2   Synthetic fibre

In recent years, chemists have developed a very strong, fire-resistant fibre called aramid fibre. It is quite like nylon, but stronger, harder to melt and more difficult to burn (figure 15). Aramid fibre is so strong it can be used to make bullet-proof vests. The photograph shows how fire-resistant it is. It is also

Figure 13   *Firemen's clothes have to protect them from fire, heat and water.*

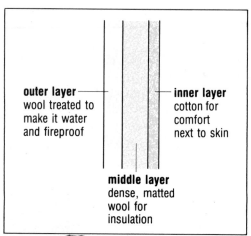

Figure 14   *Three-layer construction of a lightweight woollen fireman's jacket*

outer layer
wool treated to make it water and fireproof

inner layer
cotton for comfort next to skin

middle layer
dense, matted wool for insulation

Figure 15   *Fire-resistant material being tested in a laboratory.*

water-resistant and comfortable to wear, so it is excellent for making firemen's clothes.

The table below gives some information about three fire-resistant fabrics. Fire-resistant cotton and fire-resistant wool are natural fibres, specially treated to make them fire-resistant. Aramid is a synthetic fibre specially designed to be fire-resistant.

1 You are responsible for supplying protective clothing for your local fire brigade. The safety of your firemen is vital, but you also have to consider cost. Which fabric would you choose, and why?
2 What other uses, apart from firemen's clothing, can you think of for fire-resistant fibres?

Table 1

	Fire-resistant cotton	Fire-resistant wool	Synthetic aramid fibre
Fire-resistance	medium/good	good	very good
Insulation	medium/good	good	very good
Durability (how well it wears)	good	good	very good
Comfort	very good	very good	good
Cost	medium	medium/high	high

4 Disposable nappies

A good disposable nappy needs some important properties. It should be
- Absorbent – to soak up lots of liquid
- Leak-proof – so liquid can't leak out at the legs or waistband
- Comfortable– for the baby to wear
- Disposable – the nappy may be disposed of by burning or breakdown by bacteria. As much of the nappy as possible should be biodegradable.

Figure 17 shows a typical disposable nappy.

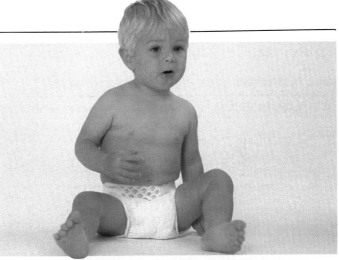

Figure 16 *There are around two million babies in Britain. About half of them wear disposable nappies - five per day on average. That makes five million disposable nappies used every day!*

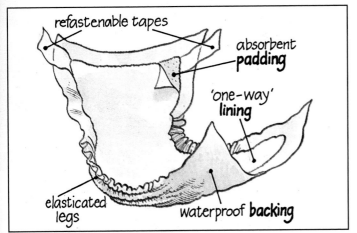

refastenable tapes
absorbent **padding**
'one-way' **lining**
elasticated legs
waterproof **backing**

Figure 17 *A typical disposable nappy*

Chemistry plays an important part in deciding the best material for making the different parts of the disposable nappy. Table 2 on the next page shows some of the materials that might be used.

Water absorption is a particularly important property. Some materials can absorb a lot of water, while some are waterproof and absorb no water at all.

Cotton and tissue paper absorb water well. They are mostly made of cellulose, and cellulose has lots of 'water hooks'.

'Superabsorbent polymer' was specially developed by chemists to absorb large amounts of water. It has large numbers of 'water hooks' and it forms a kind of jelly which holds the water under pressure.

Imagine you are a disposable nappy manufacturer. Use table 2 to answer these questions.

1 Which material or materials would you choose for the backing? Give reasons for your choice.
2 Which material or materials would you choose for the padding? Give reasons for your choice.
3 Which material or materials would you choose for the lining? Give reasons for your choice.
4 Are there any materials you would like to use but which are not in the table?

Figure 18 *This magnified photograph of tissue paper shows how the fibres are bonded together, not woven as in cloth.*

Table 2 *Materials that might be used to make disposable nappies*

Material	Description	Price	Biodegradable?	Flammable?	Waterproof?	Water absorbent?	Strength	Softness
Non-woven polypropylene fabric	see figure 18	low	no	yes	no	no	strong	soft
Non-woven rayon fabric	see figure 18	low	yes	yes	no	no	strong	soft
Woven cotton cloth	white cloth like sheets	high	yes	yes	no	fairly	strong	soft
Aluminium foil	shiny foil	medium	no	no	yes	no	strong	hard
Clear polythene sheeting	clear plastic sheet	low	no	yes	yes	no	strong	soft
Whitened polythene sheeting	opaque, white plastic sheet	low	no	yes	yes	no	strong	soft
Fluffed cellulose pulp (made from wood)	like cotton wool	low	yes	yes	no	yes	weak	soft
Superabsorbent polymer	like cotton wool	high	yes	yes	no	very	weak	soft
Tissue paper		low	yes	yes	no	yes	weak	soft

1 **Clothes**, **fabrics**, **threads** and **fibres** are all important in this chapter. Figure 19 summarises the differences between them.

2 Fibres can be natural or made artificially by chemists. Natural fibres may come from plants, like cotton. They may come from animals, like wool. Fibres made by chemists include nylon and polyester. Figure 20 sums it up.

3 There are many different types of fibres used to make clothes. Wool, cotton, acrylic, polyester and nylon are the most common fibres used. Fibres can be identified by carrying out simple laboratory tests on a piece of fabric made from the fibre. A useful test is to see how the fabric behaves when it is heated.

4 Like all substances, fibres are made up of tiny **particles**. These particles are much too small to see. The properties of a fibre (strength, stretchiness and so on) are decided by what its particles are like.

5 All fibres are polymers. **Polymers** consist of very large particles which are made up of many small particles called **monomers**. The process by which monomer particles join to a polymer particle is called **polymerisation**.

6 The main problem with cleaning clothes is removing grease and oily material. Grease does not dissolve in water. Sometimes special solvents are used instead of water – this is called **dry-cleaning**. But water *can* be used to remove grease, if detergents are added to the water.

7 For wet clothes to become dry, water must evaporate. Evaporation involves the change of a liquid to a gas or vapour. Water particles must escape from the clothes. Some fibres attract water particles more than others. Fibres that do not attract water particles well are easy to drip-dry. But they feel 'sweaty' to wear, because they do not soak up perspiration.

8 Chemists have developed ways of treating fabrics to improve their properties. Figure 21 sums up some of the treatments.

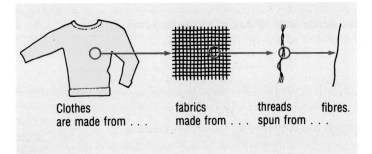

Clothes are made from . . .   fabrics made from . . .   threads spun from . . .   fibres.

Figure 19   *Clothes, fabrics, threads and fibres*

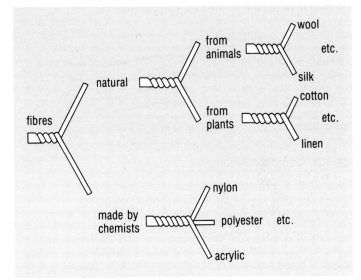

fibres — natural — from animals — wool, silk, etc.
— from plants — cotton, linen, etc.
— made by chemists — nylon, polyester, acrylic, etc.

Figure 20   *Different types of fibres*

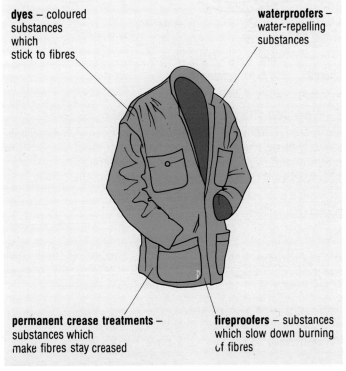

dyes – coloured substances which stick to fibres

waterproofers – water-repelling substances

permanent crease treatments – substances which make fibres stay creased

fireproofers – substances which slow down burning of fibres

Figure 21   *Different fabric treatments*

1 What kind of particles make good fibres?

Clothes are made from fabrics, and fabrics are made from fibres. But fibres themselves are made up from something still smaller. Like all substances, fibres consist of tiny particles.

Fibres are long and thin, and they need to be strong. So while substances such as water and air contain particles which are roundish in shape, this would not do for fibres. Long, thin particles are needed to make long, thin fibres, as shown in figure 22.

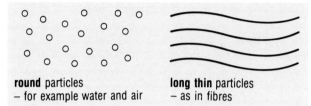

round particles
– for example water and air

long thin particles
– as in fibres

Figure 22 *Round particles and long, thin particles*

To make the fibre strong, the long, thin particles need to be lined up close to each other (figure 23). This means they can attract each other more strongly, making it more difficult to break the fibre. Just pulling the fibre helps to line up the particles.

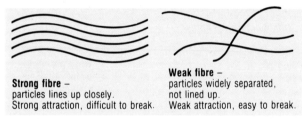

Strong fibre –
particles lines up closely.
Strong attraction, difficult to break.

Weak fibre –
particles widely separated,
not lined up.
Weak attraction, easy to break.

Figure 23 *How to get a strong fibre*

2 How can you make stretchy fibres?

Stretchy fibres like nylon or wool need to contain 'stretchy particles'. Their particles are normally closely looped up but these loops straighten out rather like a telephone cord when the fibre is pulled (figure 24).

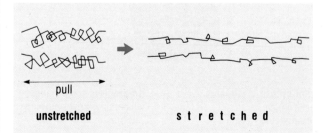

pull

unstretched

s t r e t c h e d

Figure 24 *How a fibre can be stretched*

3 What are polymers?

Fibres have long, thin particles. But how are these long thin particles themselves built up? They are made by joining lots of small units together in a chain. The small units are called **monomers**. The chain is called a **polymer**. The joining-up process is called **polymerisation**. Figure 25 explains the idea. Some polymers have more than one type of monomer unit (figure 26). Most polymers have several thousand monomer units in each chain.

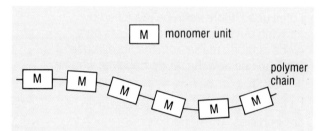

M   monomer unit

polymer
chain

Figure 25 *A polymer made from one type of monomer*

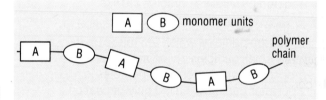

A   B   monomer units

polymer
chain

Figure 26 *Part of a polymer made from two types of monomer*

Polymers are named after their monomer. For example, the polymer made from ethene monomer units is called poly(ethene), or polythene.

You can make a model of a polymer chain by joining together paper clips (figure 27). You could call your model 'polypaperclip'!

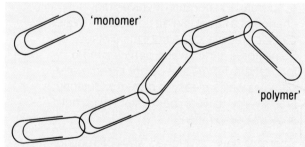

'monomer'

'polymer'

Figure 27 *'Polypaperclip'*

Many natural materials are polymers. Most of your body is made up from the polymers called **proteins**. Carbohydrates such as starch and cellulose are polymers. Polymers can also be made synthetically. Polyester and nylon are synthetic polymers.

4 What holds monomers together in a polymer chain?

Most fibres are **condensation polymers**. Their polymer chains are made by joining monomers together by condensation polymerisation. You have probably done experiments with paper strips to show how condensation polymerisation works. Figure 28 summarises the general idea.

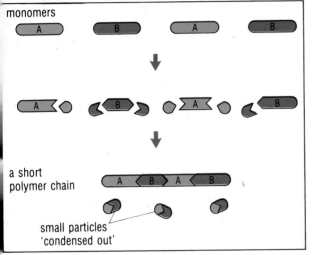

Figure 28 *Condensation polymerisation*

In condensation polymerisation, a small particle is 'condensed out' when the monomer units join together. Often, this small particle is a water particle.

For example, nylon 66 is a condensation polymer. It is made from two monomers called 1,6-diaminohexane and adipic acid. You may have made nylon by reacting these two together in the 'nylon rope trick' (figure 29). Polyesters are also condensation polymers.

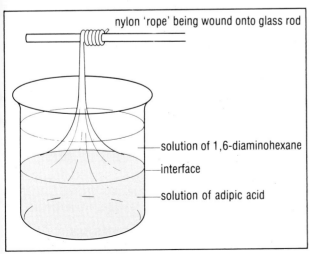

Figure 29 *The 'nylon rope trick'*

Taking it further

So far, we have talked about monomer and polymer particles without saying what these particles are made of.

Monomers and polymers are compounds. They usually contain the elements carbon and hydrogen, and sometimes oxygen and nitrogen as well. Figure 30a shows how atoms of carbon, hydrogen and oxygen are joined together in the monomers that make polyester fibre. Figure 30b shows how these two can join together to form part of a polymer chain.

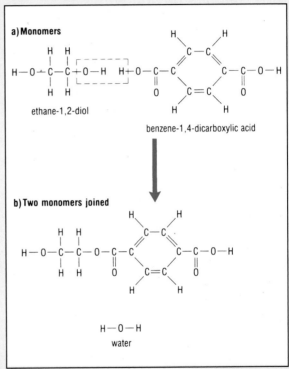

Figure 30 *Monomers join together to form part of a polymer chain.*

Suppose two more monomer units are joined onto the chain. Draw a diagram to show what the new chain would look like.

The properties of a polymer are decided by the nature of the groups of atoms it contains. For example, the groups of atoms in polyester chains tend to attract each other. The atoms on one chain attract those on a neighbouring chain. This holds the chains together quite strongly, which is why polyester makes a strong fibre.

5 Getting clothes clean

Solutions and solubility Figure 31 revises some of the key words which you may have met already to do with solutions.

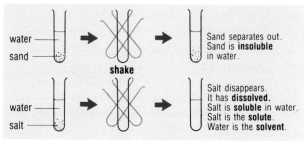

Figure 31 *Words to do with solutions*

Removing grease is the main problem when cleaning clothes. Grease and oil bind dirt onto the clothes. Grease is insoluble in water because grease and water are not attracted to each other. But grease *is* soluble in other solvents, like the solvents used in dry-cleaning.

Even though grease doesn't dissolve in water, you *can* get water to remove grease – by using detergents. Detergents dissolve in both water and grease. They work by loosening grease from clothes so that it can be washed away with water.

6 Wet clothes, dry clothes

After washing clothes we need to get them dry. Have you noticed that some clothes dry more quickly than others? This is because some fibres release the water better than others.

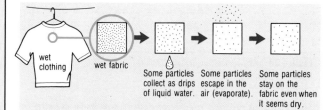

Figure 32 *How clothes dry*

What happens when clothes dry? Look at figure 32. Wet clothes have many water particles on them. Some of these go when liquid water drips off. Some of the water particles escape into the air as vapour – the water evaporates. This happens particularly quickly on a warm breezy day – a good 'drying day'.

Some fibres, like wool and cotton, attract water particles quite strongly. Their fibre particles have a kind of 'water hook' which attracts water particles. Other fibres, like nylon and polyester, attract water

particles less. This is because they have fewer 'water hooks' on the fibre particles (see figure 33).

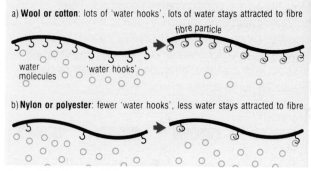

Figure 33 *Fibres with 'water hooks'*

Table 3 compares some of the advantages and disadvantages of fibres which attract water strongly, and fibres which do not.

Table 3 *Comparing the water attraction of different fibres*

	Fibres which attract water strongly	Fibres which do not attract water strongly
Examples	wool, cotton	nylon, polyester
Ease of drying	take longer to dry	dry quickly
Comfortable to wear	comfortable; do not feel 'sweaty' because they absorb perspiration	feel 'sweaty' in hot weather
Crease resistant	crease easily	crease-resistant

Taking it further

Water particles are very simple. Each particle has just two hydrogen atoms and one oxygen atom (figure 34).

Figure 34 *A water particle*

Water particles are attracted to certain other groups of atoms. The O–H and N–H groups attract water strongly. Cotton has particles with lots of O–H groups on, which is why it attracts water. Wool has lots of N–H groups on its particles, which is why wool is water-absorbent.

So 'water hooks' are not really hooks, but attractions between groups of atoms.

Things to try out

1 Using natural dyes
Try making natural dyes from strongly-coloured plant products like blackcurrants. Crush the coloured materials with warm water – you could use a food mixer. Try dyeing different materials. How effective is your dye? Does it fade? Does it wash out?

2 Tie-dyeing
You can buy excellent synthetic dyes in department stores. Try tie-dyeing a white article like a handkerchief or T- shirt. Gather a loop of the material and tie it tightly with thread. Dye the whole article, following the dye manufacturer's instructions. Where the material is tied it will not be dyed, and this can give attractive patterns.

3 How much water does wool absorb?
If you have a woollen pullover, try weighing it on a dry day and on a damp day. Is there a difference? If so, why? Do the same for a pullover made from a synthetic fibre like acrylic. How do they compare?

Things to find out

4 How is cotton produced?
5 What is linen? How is it produced?
6 Who made the first artificial fibres, and when?
7 How is soap made? And how does it remove dirt from clothes?

Making decisions

8 Choosing the right fibre
Table 4 below gives some properties of four fibres – identified here by the letters W, X, Y and Z.
   a) Which fibre would you use for each of the following jobs? Give the reasons for your choices.
      i) Making a fishing line.
      ii) Making a pullover.
      iii) Making a T-shirt.
   b) You are a manufacturer of socks. You want to make socks out of fibre X because it is warm and absorbs moisture. However, you are worried that socks made of X may wear out quickly. What could you do to get over this problem?
   c) You are a manufacturer of blouses and shirts. You want to make your garments from fibre W because it is comfortable to wear. However, its high moisture absorbency means fibre W creases easily. What could you do to get over this problem?
   d) W, X, Y and Z are actually polyester, wool, cotton and nylon, though not necessarily in that order. Which is which?

Table 4

Source: Adapted from _Take your choice_ by Norman Reid, University of Glasgow

Fibre	W	X	Y	Z
Natural or synthetic?	natural	natural	synthetic	synthetic
Strength when dry/g dec^{-1}*	3.2	1.1	4.5	4.1
Stretchiness	low	moderate	moderate	low
Moisture absorbence (%)	9	16	5	0.5
How it feels when worn	comfortable	warm and comfortable	fairly comfortable	fairly comfortable
Durability (how it stands up to wear)	good	moderate	excellent	excellent
Cost	medium	high	medium	medium

*g/dec^{-1} is a measure of force per unit area of cross-section of the fibre

9 Solvents for dry-cleaning

You run a dry-cleaning business. You have the choice of several solvents to use in your dry-cleaning machines, and you need to decide which one to use.

Table 5 below gives important properties of five solvents.

a) Decide which solvent you would use.

b) Explain your reasons for making that choice. (Look back at the section on dry-cleaning on page 49 before you start.)

Solvent	A	B	C	D	E
Ability to dissolve grease	excellent	very good	very good	excellent	excellent
Does it burn?	no	no	no	yes	no
Is its vapour poisonous?	slightly	no	no	no	no
What is its boiling point?	127°C	400°C	130°C	120°C	45°C

Points to discuss

10 Some people think it is wrong to wear clothing made from animals. Many people refuse to wear coats made from animal skins. Some people refuse to wear leather or wool. What do *you* think?

11 Some people don't like wearing synthetic fibres. They say natural fibres feel more comfortable. What do *you* think?

12 Fifty years ago practically all clothes were made from natural fibres like wool, cotton and silk. Today we wear clothes made from many different synthetic fibres as well as natural ones. What kind of clothing materials might we be wearing fifty years from *now*?

13 What matters more – the ways clothes *feel*, or the way they *look*?
a) Suppose someone invented a special spray coating that could be applied to your skin at birth. The coating is warm, waterproof and comfortable to wear. Would there be any more need for clothes?
b) Dyes do nothing to improve the comfort or efficiency of clothes. Yet nearly all clothing is dyed in some way or other. Why do people like their clothes to be dyed? Why do some people like bright colours, and some prefer dull ones? Which do *you* prefer?

Questions to answer

14 Copy out the following paragraph and fill in the missing words. Each word is used only once. The missing words are: cotton, monomers, polyester, polymer, polymerisation, polystyrene, synthetic, wool.

Fibres contain long, thin particles. These long particles are made by joining together lots of small unit particles. The smaller particles are called ___(a)___ and the long chain is called a ___(b)___. The joining up process is called ___(c)___. For example, when lots of styrene particles are joined the product is called ___(d)___. Many fibres are natural. Some come from animals, for example ___(e)___. Some come from plants, for example ___(f)___. Some fibres are not natural, but are made by chemists. They are called ___(g)___ fibres. An example is ___(h)___.

15 Jane was given two small squares of fabric, labelled L and M. She was asked to find out what fibre each was made from. She was told that one of the fibres was synthetic, and one was natural. Jane held each square of fabric in turn close to a Bunsen flame. L melted, but M did not melt.
a) Which fibre was natural and which was synthetic?
Jane now tried burning each fabric in turn. L burned with a yellow sooty flame. There was no strong smell when it burned. M was difficult to burn. When heated strongly it gave a smell of burning hair.
b) Identify L and M.
c) Why is it important for clothing manufacturers to know how fibres behave when heated?

16 a) Explain why water alone is not much use for cleaning greasy clothes.
b) Explain why adding detergent to the water helps it clean.
c) Explain why it is important to move clothes around in the water during washing.
d) Explain why hot water is usually better for washing than cold water.
e) Explain why washing must always be rinsed in clean water before drying.

17 Different clothing fibres have different properties. Four important properties are
A strength   C cost
B water absorbency   D stretchiness
Each of these properties is important for a clothing manufacturer to consider when choosing a fibre for a particular use. For each property, give *one* reason why it is important.

Introducing food

We all know how important food is in our lives. In Britain we are fortunate that there is no shortage of food. In fact we have a huge variety of foods to choose from.

energy 370kJ/88kcal
protein 3.4g
carbohydrate 14.5g
fat 2.1g
saturated fatty acid 0.4g
polyunsaturated fatty acid 1.7g

▲ **Figure 1** *Marvelfood is a new product. What does the label tell you about it? Is it something you might enjoy? Will it make you fat?*

◄ **Figure 2** *How long do you think you could last without food and water? Why does your body need food and water?*

◄ **Figure 3** *How do different foods affect your health? Are these 'health foods' really healthier?*

These questions could not be answered fully until chemists were able to analyse food accurately. They could then identify the chemicals in the food you eat and find out what happens to them in your body. The discoveries about the chemical nature of food are among the most important ever for the health and happiness of humankind.

In this chapter you will find out

◆ which of the chemicals in food are really necessary,
◆ what jobs they do in your body,
◆ about the importance of your diet.

1 Too much, too little, just right?

Whatever age you are and whatever shape or size you are, you need to eat. You need food to give you **energy** for every activity of your life. You even use up energy while you are asleep (figure 4).

> **1** Think about and then write down what your body uses energy for
>
> a) when you are asleep,
> b) while you are reading this sentence,
> c) when you are eating,
> d) when you are running.

Food also gives you the chemicals needed to **repair** the cells in your body and to make more of them as you grow.

People who have to hunt for or grow all their food are rarely overweight (figure 5). You do not see fat wild animals, though if food is scarce, you may see very thin ones (figure 6). Not only is it hard for them to get their food, they also use up a lot of energy finding it.

Millions of people in developing countries have too little to eat. This may be because of famine (not enough food can be grown) or poverty (not enough money to buy what food there is) or other more complicated reasons which you will study later in this course. The problem in the developed countries, such as the UK, is the opposite. Here, there is plenty of food in the shops and people do not use much of their own body energy to bring it home.

Figure 4
Measuring energy use during sleep

▲ **Figure 5**
Heavy physical work uses a lot of energy and keeps you slim.

◄ **Figure 6**
The leopard is using a large amount of energy to try and catch a zebra

Keeping your food intake and your weight just right is not always easy. You need to eat a **balanced** diet which must include carbohydrates, fats, proteins, minerals and vitamins. You also need to keep the correct balance between quantity and quality and this can be difficult. Many people are genuinely addicted to food – especially food containing sugar which is the most fattening and the least useful part of our food intake. The average daily consumption of sugar per person in Great Britain is 125 grams compared with 25 grams in 1880 and 5 grams in 1780 (figure 8).

Figure 7
People who eat too much too often become ill because they are overweight

Figure 8   *Average daily sugar consumption in Great Britain*

Of course the best way of keeping your weight just right is never to eat more than your body needs. Judging what your ideal weight should be is quite difficult. Health scientists think that about 30% of the adult population of Great Britain are overweight. What is even more alarming, is that about 20% of school children are also overweight. As **obesity** – being very much overweight – is linked in later life with heart disease, high blood pressure and diabetes (among other illnesses) this is bad news for the nation's health and for individual families.

2 If each member of your family eats an average amount of sugar how much sugar would you all eat in a week?

3 How much sugar does your family really use in one week?

4 Which bought foods that your family eats contain sugar?

5 Why do you think your answer to question 3 is different from your answer to question 2?

6 Suggest at least one reason why the average daily consumption of sugar has gone up 50 times in the last 200 years.

2 Slimming

If you become overweight there is only one way to get rid of the excess: eat less food than you need. To help you to do this a multi-million pound industry has developed – the Slimming Industry.

There are biscuits to eat to fill you up; there are powders to make into drinks which provide all you need for a meal; there are 'slimmers' soups, crispbreads, cereals, milk, cheese, etc, and there are hypnotists and weightwatching groups.

Look in the local supermarket or chemists and write down a few examples of slimmers foods and drinks with the 'calorie-content' of each. Make a table with these headings: Name of food, Calories/100g.

None of these slimming aids is any use unless the slimmers eat fewer calories than they need. Food chemists are working on the problem of producing tasty filling food with fewer calories. There is 'low-fat' spread which looks and tastes like margarine but is only half margarine – the other half has no energy value. The biscuits which fill you up before a meal contain methyl cellulose. This is made from the cell walls of plants or from wood pulp; it is like cotton wool. Taken with a drink of water it swells in the stomach and makes you feel less hungry even though it has no energy value because your body cannot digest cellulose.

Chemists are now working on a 'no-fat fat' which is also derived from cellulose. Their aim is to produce a substance which spreads

Figure 9 *The products of the slimming industry*

like fat, tastes like fat but has no calories in it at all! The ideal slimmer's food of the future will be an appetising meal which looks like food, tastes like food, fills you up like food but does absolutely nothing for your energy intake.

49

3 Artificial sweeteners

Some of the most useful aids to keeping food intake in balance with energy requirements are artificial sweeteners or sugar substitutes. At the moment there are three of these on the market under different trade names. They are all made by chemists and have been extensively trialled to make sure they are safe. None of them contain any 'calories'. The oldest and best-known is **saccharin** which has been used for over fifty years. During the last twenty years a number of other sugar substitutes have come on the market but some have only lasted a year or two before side-effects were reported. **Cyclamate** (sodium cyclamate), for example, gave a natural sweet taste to soft drinks but was withdrawn after a year because, given in very large doses, it was found to cause cancers in rats. **Aspartame** has been available for about ten years but may be withdrawn at any time following reports that it causes dizziness, migraine and other nerve disorders among people using it over a long period. **Acesulphame K** (K because it is the potassium salt of acesulphamic acid) has recently become available.

> If you have a slimmer in your family, look at the special diet foods they eat and list the names and trade names of the artificial sweeteners used.

The chemicals in figure 10 all taste sweet.

But why do some chemicals taste sweet, some taste bitter and others have no taste? The sensation of sweetness is thought to be because there is some sort of reaction between molecules of the chemicals and parts of the surface of the tongue, known as **receptor sites**. The process is very fast and is reversible.

The way the atoms are arranged in the molecules is the key to the process. The symbols ◄ and ▥ in the formula for aspartame are used to show the three-dimensional form of the molecule. ◄ means the atoms stick up above the page, ▥ indicates that the atoms lie below the page. Another form of aspartame has these reversed and this compound tastes bitter. The shape of the molecules makes them behave differently.

If you need to lose weight or want to make sure you stay just right, it is probably safer and certainly cheaper to give up extra sugar and sweet foods altogether.

Figure 10 *Chemical sweeteners* ▼

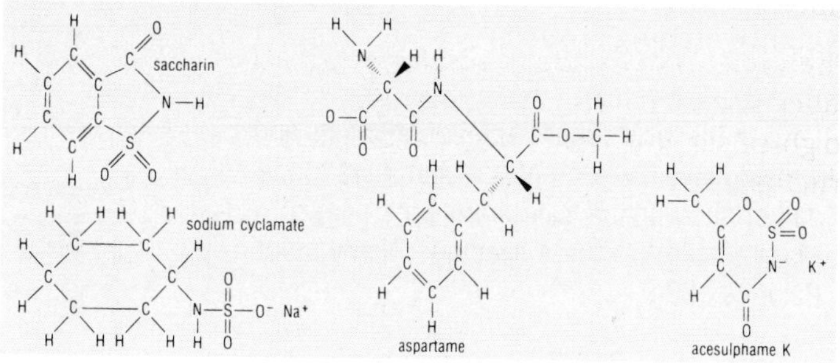

4 Special diets for special people

Diabetic foods

People who suffer from diabetes cannot produce enough insulin. Insulin is a protein hormone which regulates how carbohydrates, such as starch, decompose into smaller molecules including glucose which in turn are used to make energy. Without insulin, the glucose molecules cannot get into the cells of muscle and liver where they are stored for later use. Therefore without insulin, glucose builds up in the blood and is passed out in the urine. The cells do not have enough food and have to feed on the amino acids already in the cells which are normally needed for growth and cell repair. People who have severe diabetes must be given insulin or they will die.

Figure 11 *A young diabetic injecting herself with insulin*

The insulin given to diabetics has usually been extracted from the pancreas of pigs or cattle. These insulins differ slightly from human insulin. Recently, techniques of genetic engineering have been used to program a bacterium so that it makes human insulin. This insulin is now made on a commercial scale.

One major landmark in the development of this story occurred in 1956 when Fred Sanger (now Sir Frederick) worked out in Cambridge the sequence of amino acids in the insulin molecule. Another occurred in 1969 when Dorothy Hodgkin and her colleagues in Oxford discovered the shape into which the chain of amino acids is folded. These two discoveries have led to a clearer understanding of how insulin works in the body.

Many diabetics do not have to take extra insulin as their bodies do produce some. But all people with diabetes have to be very careful about their diet. They must keep a balance between all the essential foods (carbohydrates, proteins, fats, vitamins and minerals) while strictly limiting their glucose intake.

The sugar used in drinks and for cooking is called **sucrose**. In the body it breaks down to glucose and another sugar called **fructose**.

sucrose + water → glucose + fructose

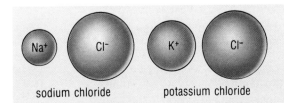

Figure 12
The structures of glucose and fructose

The chemical formulas of glucose and fructose are the same; they are both $C_6H_{12}O_6$. However the way the atoms are joined together is different.

Our bodies recognise the difference. Glucose and fructose are carried into the cells by different mechanisms. This has led to the idea that diabetics can use fructose as a safe alternative to glucose. However, some nutritional experts are cautious about the use of fructose and artificial sweeteners may be safer.

Gluten-free foods

Some babies and children are not able to digest one of the proteins in wheat. This protein, called **gluten**, causes stomach upsets and diarrhoea which can be very dangerous for small children. Children with this illness (**coeliac** disease) cannot eat anything made with ordinary (wheat) flour. All baking for them has to be done with rice flour or cornflour.

> 1 Imagine that you had to prepare the food for a coeliac friend for one day. Write down what you would give them to provide a balanced diet.
> 2 Examine a range of items in your store cupboard at home and list those which would be suitable for someone with coeliac disease.

Low-salt diets

We need sodium ions (Na^+) in the body to help move water molecules across cell membranes. This occurs as part of many processes in the body, for example in sweating. Our kidneys regulate the amount of sodium ions in the body. Excess salt is passed out in the urine.

People with heart disease or high blood pressure may be told not to add any extra salt to their food. Table salt (common salt) contains sodium ions (Na^+) and chloride ions (Cl^-).

Because many people find food without salt tasteless, people on low-salt or salt-free diets use 'salt-free salt'. This tastes something like common salt and contains potassium chloride. So it has potassium ions (K^+) and chloride ions (Cl^-). Sodium and potassium ions are transported across cell membranes by different mechanisms and they play different roles in the cell. So potassium ions do not cause the same problems as having too many sodium ions.

sodium chloride   potassium chloride   **Figure 13**

51

Baby's first cereals

For the first few months after birth a baby needs nothing but milk (figure 14). Milk provides all the essential chemicals at this age. Once a baby starts to have solid foods it is important to know what is in the foods.

Packets of baby cereals are all carefully labelled to show not only what *is* present (like most packaged food nowadays) but also what is *not* in them (figure 15). These are:

'sugar-free' to try to avoid the baby developing a taste for sweet foods,
'gluten-free' to avoid developing coeliac disease, and
'salt-free' to avoid strain on the baby's growing kidneys.

Figure 14 *Breast feeding a small baby provides the best nourishment during the first few months of life.*

Figure 15 *Packet of baby cereal showing 'food free' strips* ▶

In brief
Food

1 Your body and all the substances you eat as food are mixtures of chemical compounds. The compounds can contain many different elements but the most common are carbon, hydrogen, oxygen and nitrogen.

2 Foods contain stored chemical energy. They can burn in air to release carbon dioxide, water and heat energy.

3 If you could break down substances into smaller and smaller bits, then eventually you would not be able to break them into anything smaller. These particles are called molecules.

4 When you eat, molecules from your food are broken down by a carefully controlled sequence of reactions involving oxygen. The process is known as respiration. Many chemicals are made and energy is released. Some of the energy is used to build new molecules which act as stores of chemical energy needed to help you to move and grow. Some is released as heat which keeps you warm. The final products of respiration are carbon dioxide and water which you breathe out.

5 A reaction which gives out heat energy is described as **exothermic**.

6 Fats can provide more than twice as much energy as the same mass of carbohydrates or proteins. Unused fats are stored in the body.

7 Simple chemical tests can be carried out to identify proteins, sugars, starches and fats in foods. Tests for pH can show that foods are mildly acidic or neutral or mildly alkaline.

8 The compounds which you must eat to keep healthy are known as **nutrients**. There are seven different types of nutrients.

Essential Nutrients

Carbohydrates (starches and sugars): used to provide energy.

Fats and oils (lipids): stored in the body to provide energy when needed.

Vitamins: protect us from diseases and help the body to make use of other nutrients. Only very small quantities are needed.

Fibre (roughage): not absorbed by the body. Helps to dispose of waste material.

Mineral salts: like vitamins, are required in small quantities. They provide important elements to make more complicated molecules and for other jobs in the body.

Water: our bodies are about 75% water. Almost all the chemical processes which keep us alive and active take place in water in our bodies.

Proteins: used mainly for growth and repair but can provide energy when needed.

1 What's in food?

In food there are chemicals which you must eat to keep healthy. Other chemicals are added to make food taste and look more interesting and also to make sure the food keeps fresh. You will find out more about these later in your course.

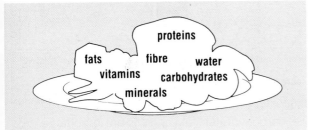

Figure 16 *Foods contain seven main groups of chemicals – **nutrients***

Although all the food types in figure 16 are important you need to know about three of them – proteins, carbohydrates and fats – in more detail.

Proteins
The cells and tissues in your body are made of **proteins**. During your life, your body keeps using proteins to replace and repair damaged or dead cells.

All proteins are built from smaller molecules called **amino acids**. These contain carbon, hydrogen, oxygen and nitrogen atoms. Some amino acids, like methionine (figure 17), also contain sulphur atoms.

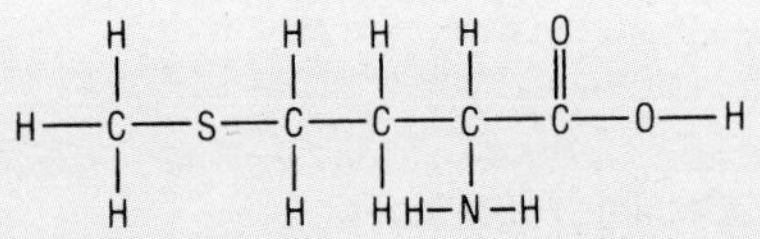

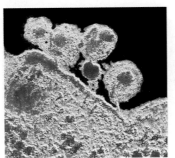

Figure 17 *Chemical formula of an amino acid – methionine*

Figure 18 shows a small piece of a protein molecule. This piece is made from five amino acid molecules. The amino acids are shown by different shapes. The sequence in which the amino acids are joined is special to each protein.

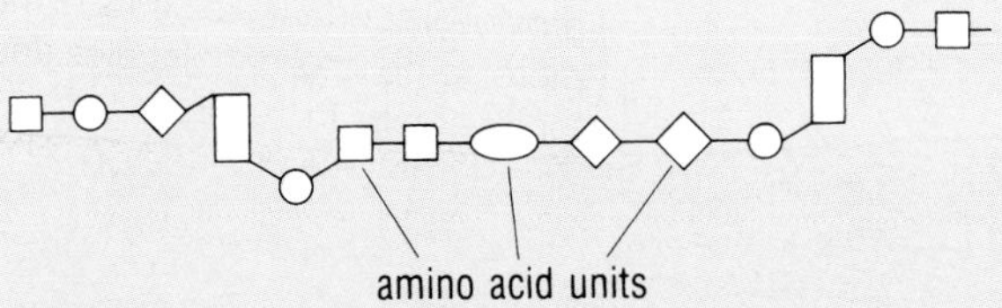

Figure 18 *A piece of a protein chain formed by five different amino acids*

All proteins are made from different sequences of amino acids. With twenty amino acids to choose from, thousands of proteins can be made. They include **haemoglobin**, which is the red colouring-matter in blood, and **keratin** in hair. Haemoglobin is a medium-sized protein molecule, containing about 600 amino acid units. Some proteins contain more than 4000 amino acid units.

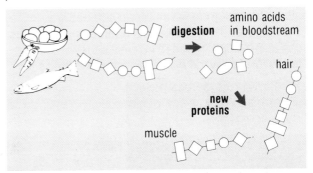

Figure 19 *The AIDS virus (the purple coloured blobs in this electron micrograph) is shown here breaking from an infected cell. The envelope surrounding the virus is made of protein.*

Proteins in the food you eat are broken down to amino acids in your digestive system. The amino acids are carried around your body by your bloodstream to your body cells. Then they join together in various combinations to make new protein chains (figure 20).

Figure 20

Your body is able to convert one amino acid to another depending on which ones are needed most. But there are eight of the total of 20 amino acid types which it cannot make. If you are to stay healthy, your food must contain proteins which break down to give these eight amino acids. They are called the **essential amino acids**. The richest sources of food proteins which give the essential amino acids are meat, fish, milk, cereals and vegetables, such as peas and beans.

Carbohydrates
Carbohydrates in the form of starches and sugars, are the main source of energy for your body. They are not used for growth or repair. All carbohydrates contain carbon, hydrogen and oxygen.

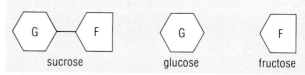

Figure 21
Bread and wheat ▶

▲ **Figure 22**
Rice plant

Figure 23
Rice grains ▶

One important source of starch is cereals – wheat, oats, rice, etc. – from which bread and pasta are made (figures 21 – 23). Another source is vegetables, such as potatoes. These plants make their carbohydrates by the process of **photosynthesis**.

Figure 24
Sugar cane ▼

▲ **Figure 25**
A selection of sugars produced from sugar cane

Sugars are the most well-known carbohydrates. To a chemist, **sugars** is the name for a family of substances, all of which have similar properties. The sugar you use to sweeten your food is one of this family (figures 24 and 25). It has the chemical name **sucrose** and formula $C_{12}H_{22}O_{11}$.

Sucrose can be broken down in the body to form two smaller sugars. One is called **glucose** and the other **fructose** (figure 26). They both have the same chemical formula $C_6H_{12}O_6$ and yet they are different substances. Glucose contains a ring of six atoms joined together. Fructose contains a ring of five atoms.

```
   G   F        G        F
  sucrose    glucose   fructose
```

Figure 26

Fats (also known as **lipids**)
Like carbohydrates, fats and oils are important sources of energy for your body. An oil is a fat which is liquid at room temperature.

You eat fats in all sorts of foods (figure 27). There is fat in meat, milk, butter and cheese. These are '**animal fats**'. Many plant seeds and nuts are rich in fats, usually in the form of oil. Sunflower seeds, soya beans, peanuts and olives all provide you with '**vegetable fats**'. To many peoples' surprise, peanuts are about 50% fat.

Figure 27   *A selection of fatty foods*

Like carbohydrates, fats contain only atoms of carbon, hydrogen and oxygen. But their structures are very different from carbohydrates. Each fat molecule has a 'stem' formed from a chemical called **glycerol**. Attached to the glycerol stem are three molecules of acids known as **fatty acids**. Each of these molecules has a long chain of carbon atoms, with mainly hydrogen atoms attached to it.

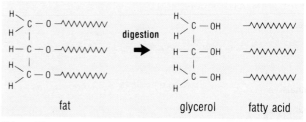

Figure 28

During digestion fats break down into glycerol and fatty acids (figure 28). These move around the body in the bloodstream. They re-combine as new fats which can be stored in the tissues below your skin.

Doctors now encourage people to eat fats which contain fatty acids with some double bonds between the carbon atoms. These fats are known as polyunsaturates. They are healthier than saturated fats, which do not have double bonds between the carbon atoms. You will find out more about this later in your course.

2 How do you keep warm?

Have you measured your body temperature with a thermometer? A healthy person should find that their normal temperature is about 37 °C. Except in extremely hot conditions, 37 °C is higher than the air temperature around you. Something inside you is producing heat energy which makes your body warm. Energy is also needed to keep your muscles active and your brain working.

So where do you get the energy from? A clue comes from analysing the air you breathe out (exhale). It is not the same as the air your breathe in (inhale), see figure 29. The main differences are shown in table 1.

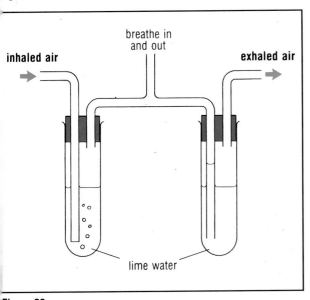

Figure 29

Table 1

	Nitrogen	Oxygen	Carbon dioxide	Water vapour	Other gases
inhaled air/%	78	21	0·03	variable	1
exhaled air/%	78	16	4	always higher	1

Look carefully at table 1 and you will see that something is happening inside your body which:

◆ involves using up some of the oxygen you breathe in,
◆ causes an increase in the carbon dioxide and water vapour in the air you breathe out and raises the temperature.

These changes must be due to the millions of chemical reactions occurring inside you all the time.

In any chemical reaction one set of chemicals (the **reactants**) changes into another set of chemicals (the **products**). During reactions there are always changes in the energy of the chemicals. Sometimes the reaction *takes in* energy from the surroundings. This makes the surroundings colder. Chemists call this type of reaction **endothermic**. More often, chemical reactions *give out* energy, making the surroundings hotter. Chemists call this type of reaction **exothermic**.

There are some endothermic reactions happening inside your body. However most reactions in your body are exothermic as you are usually hotter than your surroundings. These exothermic reactions occur between some of the food chemicals you eat and the oxygen you breath in. The products of the reactions are carbon dioxide, water and energy.

The name given to this whole chemical process is **respiration**. It is a form of **oxidation**. The oxygen you breathe in oxidises the food chemicals.

RESPIRATION:

Food chemicals + Oxygen → Carbon dioxide + Water + Energy

| Eaten | Breathed in | Breathed out |

Some of the energy from respiration is used to help you grow, move, think and do work. The rest is given out as heat and keeps you warm. So one of the main uses of food is to act as **fuel** for your body. The well known fuels – petrol in car engines and coal on fires – also react with oxygen from the air to produce carbon dioxide, water vapour and energy.

There is a simple demonstration which shows the similarity between the oxidation of food chemicals and other fuels when they burn (see figure 30 on the next page).

If some food or a fuel is burnt, carbon dioxide and water vapour are produced. In this experiment, air is drawn slowly through the apparatus, from left to right.

The presence of water in the products of combustion is shown when the anhydrous copper(II) sulphate turns from white to blue. The lime water turns cloudy to show that carbon dioxide is formed.

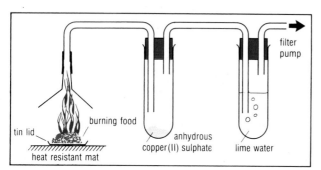

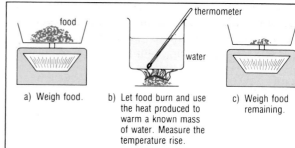

a) Weigh food.

b) Let food burn and use the heat produced to warm a known mass of water. Measure the temperature rise.

c) Weigh food remaining.

Figure 31

Then calculate as follows:

$$\text{Energy for heating the water from 1 g of food} = \frac{\text{mass of water heated (g)} \times \text{temperature rise (°C)} \times 4.2}{\text{mass of food burned (g)}}$$

For a 1g sample of **sugar** the following results were obtained:

Mass of sugar burned = 1.0 g
Mass of water heated = 200 g
Temperature rise = 20°C

$$\text{Energy released from burning 1g of sugar} = \frac{200 \times 20 \times 4.2}{1000}$$

= 16.8 kJ

When a 1g sample of **margarine** was treated in the same way, the energy released was much greater – 36.6 kJ.

Burning is a very fast set of chemical oxidation reactions between fuel and oxygen. It produces a lot of heat and light very quickly. Fortunately the exothermic reactions which occur during respiration in your body, take place more slowly. If you take heavy exercise, the chemical reactions in respiration speed up. More food is oxidised, which releases energy faster. You feel much hotter. Even so, you will not burst into flames!

3 How can you measure energy?

Energy is normally measured in units called **joules (J)**. It takes 4.2 joules to raise the temperature of 1 gram of water by 1 degree Celsius. In measuring the energy released by burning fuels, it is more convenient to use a larger unit. This is the **kilojoule (kJ)**, which is 1000 joules. A burning match releases about 4 kJ.

You may find another unit which is sometimes used for the measurement of energy from food. This is the **calorie (cal)** or **kilocalorie (kcal)**. The definition of a calorie is that 1 calorie of energy will raise the temperature of 1 gram of water by 1 degree Celsius.

1 cal = 4.2J
1 kcal = 4.2 kJ

In modern science only the joule and kilojoule are used. The calorie and kilocalorie appear on food labels and in some books and magazines about nutrition and diet.

Food scientists have found methods for measuring the amounts of energy produced from different foods. This is very difficult to do accurately in a school laboratory. A simple method is shown in figure 31.

4 What foods give you energy?

All the main food chemicals contain carbon and hydrogen plus other elements. When food chemicals react with oxygen during respiration, the exothermic reaction may be shown as:

Food

$$\begin{matrix} C \\ H \\ O \end{matrix} \ + \ O_2 \ \rightarrow \ CO_2 \ + \ H_2O \ + \ \text{Energy}$$

Carbohydrates and fats are the main foods in your body. One gram of carbohydrate releases about 17 kJ during respiration. One gram of fat releases about 37 kJ, more than twice as much. The figure for fat is the same for saturated and unsaturated fats. The low-fat spreads you can buy provide less energy for each gram you eat. This is because water is mixed with the fat!

Your stores of carbohydrate would not last long if they were your only source of energy, for example, even world class marathon runners would run out of energy about halfway through the race if they relied on carbohydrate as a fuel. You should therefore not believe the story that glucose is the main or the most important source of body energy. Nor should you believe that glucose is more readily available than fat. When you are resting, rather more than half your energy is supplied from fat. The proportion changes during activity, but always you are oxidising a mixture of fats and carbohydrates.

How can you find out what is in food?

When you are investigating food, many simple tests can be tried in a school laboratory. Only small samples of different foods are needed. Here are some examples of tests you might try.

a) **Examine the food.**
   What is its colour, texture, hardness? Is it like fibre, a jelly or some other form?

b) **Test the food's solubility in water.**
   Does it dissolve quickly or slowly or not at all?

c) **Test the food with a pH indicator.**
   When tested directly or shaken in water the food may show an acidic, alkaline or neutral nature. The pH scale of 1-14 is used for comparing acidity and alkalinity (figure 32). Universal Indicator can show where the pH of a liquid lies on the scale.

Figure 32 *Universal indicator scale – the pH of a liquid is found by matching the colour on the indicator which has been dipped in the liquid with a reference chart like the one above.*

d) **Heat the food sample gently, then more thoroughly.**
   (i) Does it melt, boil or break down (decompose)? Is any residue left after heating? A black deposit would indicate the presence of carbon in the compounds of the food.
   (ii) Are any gases given off? You should always test for the gas carbon dioxide. If it is present this shows that there are compounds containing carbon in the food.

The test for carbon dioxide is to pass the gas from the heated sample into lime water (calcium hydroxide solution) (figure 33). If the solution goes milky (a white precipitate), carbon dioxide is present.

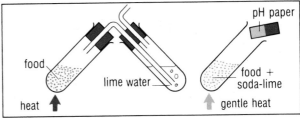

Figure 33 *Testing for carbon dioxide*

Figure 34 *Testing for nitrogen*

e) **Test for water in the food.**
   Water in the food may be present in liquids such as juices. It may also be released when the food is heated gently. Water can be detected by its effect on dry cobalt chloride paper. This changes from pale blue to pink.

f) **Testing for nitrogen in a food.**
   If nitrogen is present, all or most of it will be combined with carbon and hydrogen in proteins. However, when the food sample is heated with soda-lime, the proteins break down (figure 34). Ammonia gas is released. Ammonia contains nitrogen combined with hydrogen (NH_3). Ammonia is an alkaline gas. It turns red litmus paper blue or gives a colour on pH paper which shows a pH well above 7.

g) **Tests for the main food chemicals (nutrients).**
   There are fairly simple tests for starch, glucose, proteins and fats, as shown in figure 35.

Figure 35 *Testing for nutrients*

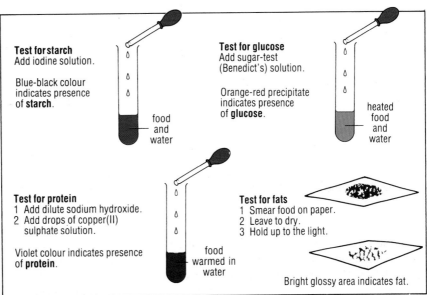

Test for starch
Add iodine solution.

Blue-black colour indicates presence of **starch**.

Test for glucose
Add sugar-test (Benedict's) solution.

Orange-red precipitate indicates presence of **glucose**.

Test for protein
1 Add dilute sodium hydroxide.
2 Add drops of copper(II) sulphate solution.

Violet colour indicates presence of **protein**.

Test for fats
1 Smear food on paper.
2 Leave to dry.
3 Hold up to the light.

Bright glossy area indicates fat.

Things to try out

1 *How would you detect the presence of carbon dioxide in breathed-out air?*
Design experiments to investigate if the amount of carbon dioxide breathed out is different before and after eating a meal.

2 *Investigating high-fibre breakfast foods.* One of the properties of fibre (roughage) in your diet is that it absorbs water and swells.
   a) Design experiments to compare the water-absorbing properties of different high-fibre breakfast foods.
   b) Which breakfast food is the most effective absorber of liquids?
   c) Which swells the most?

3 *Indigestion pains are said to be caused by too much acid in the stomach.*
   a) Investigate the anti-acid medicines which you can buy in shops and are used to reduce stomach acidity.
   b) What are the active ingredients in the anti-acids?
   c) Try adding crushed anti-acid tablets to an acidic substance such as lemon juice or vinegar. Does the acidity disappear?
   d) Design experiments to find out which anti-acid is the most effective.

Things to find out

4 a) What is the modern chemist's name for 'bicarbonate of soda'?
   b) Why can it be used as a raising agent in the baking of bread or cakes? What are the chemical reactions involved?

5 How is sugar extracted from sugar beet on a large scale? Are you able to obtain sugar crystals from a sugar beet in your laboratory?

Points to discuss

6 What sort of foods should be packed in survival kits for storing in aeroplanes or lifeboats? The people who use them may have to live without any means of cooking for about a week, in very difficult conditions, such as in an open boat or in the desert.

7 How would you convince people in your school that there are good reasons for eating food which is high in fibre and low in fats?

8 Someone says 'Food is just a lot of chemicals we happen to eat. It would be much easier and healthier if we could get all our food as a few pills every meal.' What do you think about this idea?

Questions to answer

9 The labels on five jars containing white powders in a food cupboard have fallen off and become mixed up. The labels say

protein extract   starch   glucose   salt   baking soda

How would you find out which substance is in each jar?

10 A TV advertisement for a drink claims that 'it gives you energy'. A bottle containing 1 kg of the drink costs 99 pence. On the label of the bottle you read that the liquid contains 'glucose 20%, citric acid 0.5%, lactic acid 0.1%, vitamin C 0.2% and carbonated water'.
   a) What does 'carbonated water' mean? How would you test the drink to prove your answer?
   b) What is the chemical source of energy in the drink?
   c) What mass of this chemical is there in the bottle?
   d) How much would the same mass of this chemical cost if you bought it in pure form from a shop?
   e) What is your opinion about the value of this drink as a source of energy?

11 The energy we use in various activities has been measured.

Activity	Energy used/ kilojoules per minute	Activity	Energy used/ kilojoules per minute
sleeping	4	cycling	25
sitting	6	dancing	25
standing	7	running	35
walking slowly	13	football	35
walking quickly	21	swimming	35

You will also have seen tables of data about the energy provided by different foods.
   a) From the time you get out of bed until lunchtime, calculate approximately how much energy you have used.
   b) Calculate how much energy you gained from your breakfast.
   c) Decide if you are eating too much or too little of the right type of food at breakfast.
   d) One cream meringue provides about 850 kJ. To use up just this energy, for how long would you have to
   (i) sit still,   (ii) walk quickly,
   (iii) take vigorous exercise?

Introducing transporting chemicals

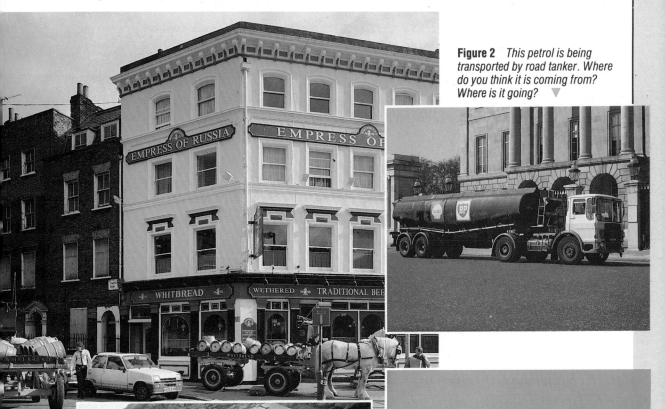

Figure 2 *This petrol is being transported by road tanker. Where do you think it is coming from? Where is it going?* ▼

▲ **Figure 1**
Beer is usually transported by road tanker from a brewery to a distribution depot for pubs and hotels. But this beer is still transported by horse and cart.

Figure 3 *Tankers carrying hazardous chemicals must display a hazard warning sign.*

Figure 4 *This ship is transporting liquid sulphur which will be used to make sulphuric acid. Why is it being transported by ship?*

Petrol, beer and sulphuric acid are *all chemicals* which have been made from raw materials. In the photographs above they are all being transported from one part of the country to another. Sulphuric acid is different from the other two chemicals in that you are very unlikely to use it directly. It is on its way to a factory where it will be used to make something else, such as paint or detergent, which you will then use. What other differences can you think of between these three chemicals?

In this chapter you will see

◆ why chemicals are transported around the country and in what ways chemicals can differ from each other,
◆ the different methods which are used to transport chemicals and what can be done to deal with chemical spillages,
◆ the shorthand system of indicating the hazards on chemical tankers and containers which must be understood by members of the fire, police and medical services,
◆ the shorthand system of representing chemical substances which is used throughout the world.

1 What is the chemical industry?

Some people work in the **service industries**. This means they provide a service to other people. A hairdresser, a refuse collector, a probation officer, a doctor, a shop assistant, and your teacher all provide services to other people.

To provide these services they each need certain materials to assist them in their work. For example, they probably all use paper but they will also use other things needed for their particular job.

> **1** Spend a few minutes writing down the materials that some of these people need.

All these essential materials are produced by the **manufacturing industries**. These industries also produce all the goods you use at home and at school – for work and leisure.

The **chemical industry** is one of the most important and successful parts of this manufacturing industry. It takes **raw materials** – oil, gas, coal, minerals, air, water – and makes a wide range of chemicals from them. These chemicals are used to make anything from bread to compact discs (figure 5). The chemicals it makes are exported to other countries around the world as well as being used here.

Inside a chemical works, some chemicals are produced in very large quantities. Sometimes, as a result of the process, other chemical products are made for which no use can be found. These are called **waste products**.

The part of the works which produces a chemical is called a **plant**. Some of the chemical reactions which take place in a plant occur at high temperatures so that a source of heat energy is needed. Also a lot of electrical energy is used moving substances from one part of the plant to another.

People with different skills are needed in the chemical industry. You can get a picture of what happens inside the 'box' in figure 6 by thinking about what these people do.

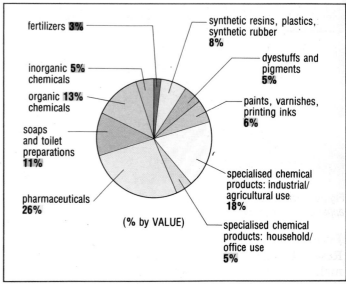

Figure 5 *The pie chart shows the range of products made by the chemical industry in Britain and their relative value.*

Figure 6 *You can think of a particular chemical industry as a large box - things go into the box, things happen inside the box, things come out of the box.*

> **2** Draw a box on a piece of paper and, using what you have read so far, start to develop a rough diagram of the chemical industry showing what goes in and what comes out.

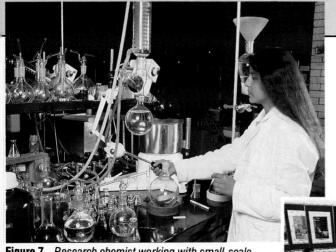

Figure 7 *Research chemist working with small-scale apparatus in a laboratory.*

People in the chemical industry

Research chemists find new processes and make new products (figure 7). They work closely with the people in the **marketing** and **sales departments** who will be able to say if the product is wanted. The marketing department will also advise on how the product ought to be changed so that there will be even more demand for it. At this stage the process will be operating in a **laboratory**.

The product will be given to customers for trial and the **financial experts** will estimate how much the product will cost to make.

The chemical engineers then have to scale up the process and design a full-scale plant like the one in figure 6. Plants such as these can cost up to 150 million pounds to build. The research and development costs can also be high, particularly for high-value products, like medicines. Some cost as much as 15 million pounds before production can begin.

Sometimes the product from a plant might go directly on sale to the public but often the chemicals produced have to be transported to another branch of manufacturing industry where they are used to make other products. **Transport workers** bring materials in and out of the factory.

There are **managers** who control the whole operation. There are also people who look after the needs of the people working there – **canteen staff**, **medical staff**, **training officers**, **safety officers** and so on. So the chemical industry has its own service departments within it.

Figure 8 *When the plant is running there will be people operating it and production chemists in charge. This is the control room of a chemical plant.*

Figure 9 *The sales department, seen here having a conference, will be responsible for selling a product.*

Figure 11 *Transport workers bring materials in and out of the factory.*

Figure 10 *Maintenance workers are needed to keep the plant running.*

3 Use the information on this page to complete your diagram of the chemical industry started in question 2.

During your chemistry course industrial processes will be mentioned many times. Sometimes only the chemical reactions being used and the useful products will be mentioned but remember that all these other activities are essential in the complete manufacturing process.

2 Who first thought of atoms?

1 Write a sentence explaining what you think an atom is.

Everyone, at least by the time they are old enough to attend secondary school, has heard of **atoms**. Some of your early ideas about atoms will have come from TV or the newspapers. The atomic bomb, more often called a nuclear weapon these days, is what many people think of when the word atom is used.

It is important that at this stage you sort out in your mind a reasonable picture of what atoms are. As you learn more about science you will be able to add more detail to this picture.

So, what is the story of the atom? Someone had to be the first person to think of the idea of atoms. Atoms are too small to be seen so they have not always been known about (figure 13).

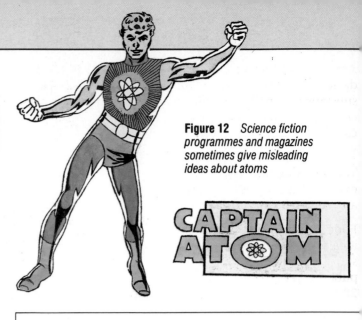

Figure 12 *Science fiction programmes and magazines sometimes give misleading ideas about atoms*

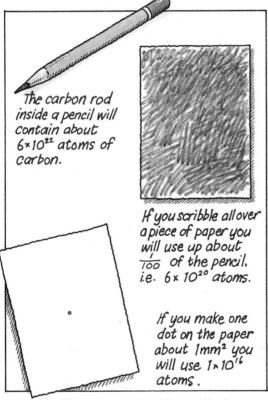

The carbon rod inside a pencil will contain about 6×10^{22} atoms of carbon.

If you scribble all over a piece of paper you will use up about $\frac{1}{100}$ of the pencil, i.e. 6×10^{20} atoms.

If you make one dot on the paper about 1mm² you will use 1×10^{16} atoms.

Figure 13 *The total population of the world is about 4 000 000 million (4×10^{12}). This means there are 2500 times more atoms in a pencil mark of 1 mm² than there are people in the world.*

Figure 14 *Democratus and his colleagues searching for the atom.*

Around 400 BC some Greek philosophers were discussing the idea that everything consisted of tiny indivisible particles. A philosopher is a person who likes to consider questions or develop ideas and arguments in order to have more knowledge and understanding of the world. As part of the discussion Democratus, one of the philosophers, put forward this question:

'Is it possible to keep on cutting up a piece of a substance, such as gold, into smaller and smaller pieces or would you reach a point where the particle left was the smallest possible piece of gold?'

After considering this question Democratus decided that there must be, for every substance, a smallest possible particle which could not be divided again. This particle came to be called an atom, from the Greek work *atomos* which means 'indivisible'.

The Greek philosophers did not do experiments so Democratus did not have any evidence to support this idea.

For centuries after these early Greeks the idea of atoms was dropped. Then a major step forward came when John Dalton published his Atomic Theory 2000 years later in 1813.

It is not clear how he first came to think of his theory but what he wrote in his note book in 1803 shows that he thought:

◆ **matter** (which means everything) consists of small particles (which he called **atoms**),
◆ atoms cannot be divided up,
◆ all atoms of an **element** (a single substance like gold) are identical.

He also said that atoms could not be created or destroyed. He was right! Just think – the atoms you are made of existed long before Democratus was discussing his ideas! The same atoms could have been part of Julius Caesar, Queen Elizabeth I, a beetle and a dog.

Figure 15   *John Dalton was born in a small village in Cumbria. He started his own school in the village when he was only 12. Eventually he became a teacher in a college in Manchester and it was there he did most of his research.*

Dalton was an experimental scientist and so unlike the Greek philosophers he put his theory forward to explain the results of experiments he had carried out. In these experiments the masses of different elements combining together were measured.

In some ways Dalton's theory has had to be modified in more recent years but the basic idea is still accepted:

◆an element consists of tiny particles which we call atoms,
◆the atoms of one element, say gold, are different from the atoms of all other elements.

2 Look back to the answer you gave to question 1 and see how you need to change your answer or add to it now that you have read these two pages.
3 The basic ideas of Dalton's theory gives us a definition of an atom. How could this be turned round so that it is used to define an element?

1 Everything we use is manufactured from raw materials.

All of the raw materials available to us on this planet can be divided into three groups: living, non-living, previously living

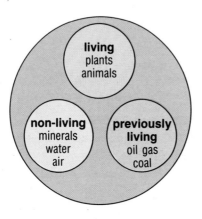

living
plants
animals

non-living
minerals
water
air

previously living
oil gas
coal

2 The chemical industry is a large and important part of the manufacturing industry. Sometimes it converts the raw materials into products which we buy, such as oil into petrol. But frequently it first converts the raw materials into chemicals which are then used to make products we buy.

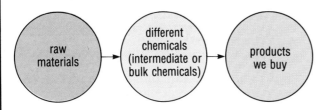

raw materials → different chemicals (intermediate or bulk chemicals) → products we buy

3 Many of these intermediate or bulk chemicals are chemicals which are used in your laboratories.

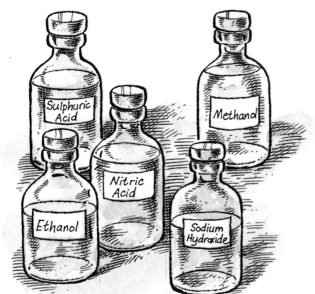

Sulphuric Acid
Methanol
Nitric Acid
Ethanol
Sodium Hydroxide

Figure 16 *Common laboratory chemicals*

4 One factory might produce a very large quantity of say sodium hydroxide which will then be used to make a wide range of products. Some of these products may be made at the same factory, or one nearby, but others will be made at factories in other parts of the country. This is why it is necessary to transport chemicals (page 66).

5 Bulk chemicals can be transported by:

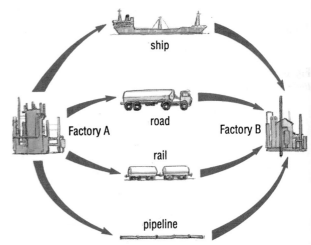

ship

Factory A road Factory B

rail

pipeline

Figure 17 *Methods of transporting chemicals*

The method used depends on geographical, economic, social and environmental factors.

6 Different chemicals have different properties. If you did an investigation into how each of the chemicals mentioned in point 3 reacts with water, fabric, leather (skin) and metals you would see some of these differences. Also some of the chemicals are flammable and some are not.

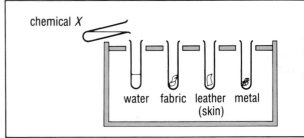

chemical *X*

water fabric leather metal
(skin)

Figure 18 *Testing an unknown chemical*

Because all these widely different chemicals are often transported around the country, it is necessary to have a system of labelling their containers which can be understood by everyone involved. For example, if there is an accident involving a chemical the police and fire service need to know if it is corrosive, if it is poisonous, if it reacts with water, if it is flammable and in particular how to deal with a spillage.

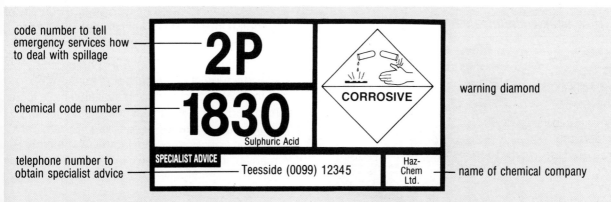

code number to tell emergency services how to deal with spillage

chemical code number

telephone number to obtain specialist advice

warning diamond

name of chemical company

Figure 19 *The Hazchem (Hazardous Chemicals) code is understood by all emergency services.*

The emergency services can work out from the Hazchem labelling code (figure 19):

◆ if the chemical is likely to react violently,
◆ whether to wear full protective clothing or breathing apparatus,
◆ whether the chemical can be washed away down the drains or must be kept out of the drains (contained),
◆ whether or not they should evacuate people from the surrounding area.

7 Some of the factors considered when deciding where to build factories are shown in figure 20.

Figure 20 *Factors affecting the siting of a chemical factory*

8 An **element** is a substance which cannot be broken down into anything simpler. The smallest particle of an element is called an **atom**. Atoms of one element are different from atoms of all other elements. Each element is represented by a **symbol**, which is a shorthand for the element (page 10).

The atoms of some gaseous elements are joined together in pairs so that chlorine gas is represented by Cl_2 rather than Cl. Similarly oxygen is O_2, nitrogen N_2 and hydrogen H_2.

Compounds contain the atoms of more than one element joined together (not just mixed up). Each pure compound can be represented by a **formula** which is a shorthand for the compound. For example, $CaCO_3$ represents calcium carbonate and tells you that it contains the elements calcium, carbon and oxygen atoms in the ratio of 1 : 1 : 3.

The smallest particles of some compounds and elements are called **molecules**. For example, H_2O represents a molecule of water and H_2 a molecule of hydrogen.

9 Chemical reactions between elements or compounds can be represented by **word equations** which summarise the reactions. They list the starting substances (reactants) and the products. For example, the burning of natural gas (methane) can be represented by the word equation:

methane + oxygen → carbon dioxide + water

Alternatively, an equation which is made up of symbols and formula can be used:

$$CH_4 + 2O_2 \rightarrow CO_2 + 2H_2O$$

This is called a **balanced equation**.

Symbols, formulas and balanced equations are the international language of chemists and are used by chemistry students and chemists all over the world.

1   How should chemicals be transported?

*Transporting you (**figure 21**) and your letters and parcels (**figure 22**) is expensive. Transporting chemicals is both expensive and potentially hazardous.*

If you see a road tanker transporting sodium hydroxide then it is probably being taken from one factory (where it was made) to another where it will be used to make something else. The manufacturer who is selling this chemical has decided to use road tankers rather than trains, ships or pipelines. Only small quantities of specialised chemicals would be sent by plane.

In the end such a decision is based on costs but this does not necessarily mean that the cheapest method is always chosen. For example, it may be better for environmental reasons, to build a costly new rail line to a factory rather than transport the chemical by tanker on existing roads. In this case the cost of the rail link would have to be calculated and then be considered alongside the benefit to the environment.

When making a decision about transporting any chemical, the type of chemical being transported and the methods of transport available have to be considered, but in particular cases other factors may also be important. For example, if chemical X needs to be transported from factory A to factory B, then many of the factors shown in figure 23 could be important.

In predicting the cost of a method of transport, the manufacturer needs to consider both capital costs and running costs (figure 24).

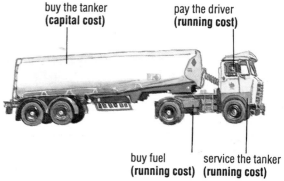

Figure 24   *Some of the capital costs and running costs for road transport*

A pipeline to transport chemicals might be very expensive to build (**capital cost**) but cheap to maintain (**running cost**) (see figure 25).

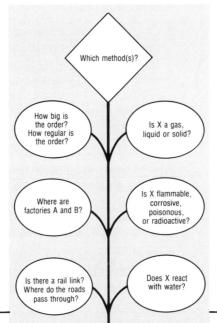

Figure 23   *Factors affecting the method chosen to transport chemical X from factory A to factory B.* ▶

Which method(s)?

How big is the order? How regular is the order?

Is X a gas, liquid or solid?

Where are factories A and B?

Is X flammable, corrosive, poisonous, or radioactive?

Is there a rail link? Where do the roads pass through?

Does X react with water?

THE DECISION

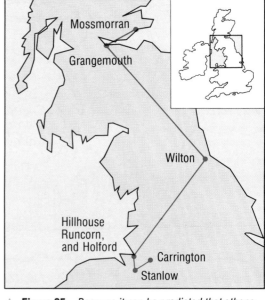

▲ **Figure 25**   *Because it can be predicted that ethene will need to be transported between the places shown, on a regular basis for a number of years, it is worthwhile building a pipeline.*

2 How do chemicals differ?

This is an impossible question to answer! There are millions of known chemicals and there are many, many ways in which they differ. However, it is important for you to know about a few of the ways in which they can differ. At this stage you do not have to remember exactly how one particular chemical differs from another.

Some differences are physical differences. For example, figure 26 shows three substances at room temperature and pressure, which are each in a different physical state.

Figure 26

Another difference is that some substances will mix with water and others will not. Some of those that do mix react with the water to form new substances, but others just **dissolve** (the particles of the substance mix up with the particles of water) (figure 27).

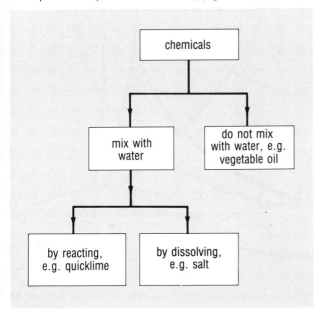

Figure 27

If one substance reacts with water and the other doesn't then this is a **chemical difference** rather than a physical one. Another chemical difference is that some substances are flammable and others are not (figure 28).

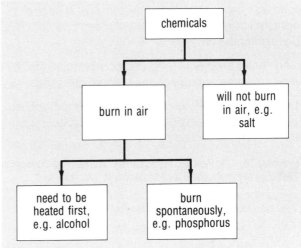

Figure 28

Some chemicals corrode skin or fabric and others do not. A lot of the substances which are corrosive are either acids or alkalis (figure 29). This is also a chemical difference.

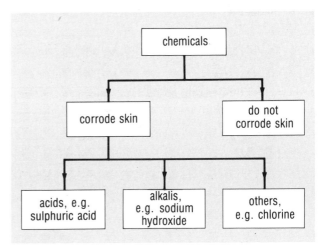

Figure 29

You can usually tell that a substance is an acid from its name and later you may be able to recognise some alkalis from their names. But, if you are not sure, a simple test is to use **litmus paper**. This is paper which has been soaked in a dye. When the paper is dipped into:

an acid it turns from ▨ to ▨

an alkali it turns from ▨ to ▨

3 Elements, compounds, symbols and formulas

People in the emergency services, who do not usually have any specialised chemical knowledge, can understand the Hazchem Code (see page 65). But chemists and chemistry students all over the world also have a shorthand system that is used to represent *all* pure chemical substances and conveys a lot of information.

Every material object, living and non-living, in the whole universe is made of one or more elements.

There are about 100 elements. The arrangement of them in figure 30 is called the **Periodic Table**. Each element is represented by a symbol.

Each element consists of tiny particles called **atoms**. The atoms of one element, for example copper, differ from the atoms of all other elements.

The atoms of different elements can combine to form substances which are called **compounds**. The smallest parts of all compounds are called **molecules**.

Compounds are represented by **formulas**. The formula is made up of the symbols of the elements which are in the compound and small numbers which show in what ratio the atoms of the elements are present. Examples are shown in red in figure 30.

▼ **Figure 30** *The Periodic Table is a way of arranging the elements which together make up all the substances in the universe.*

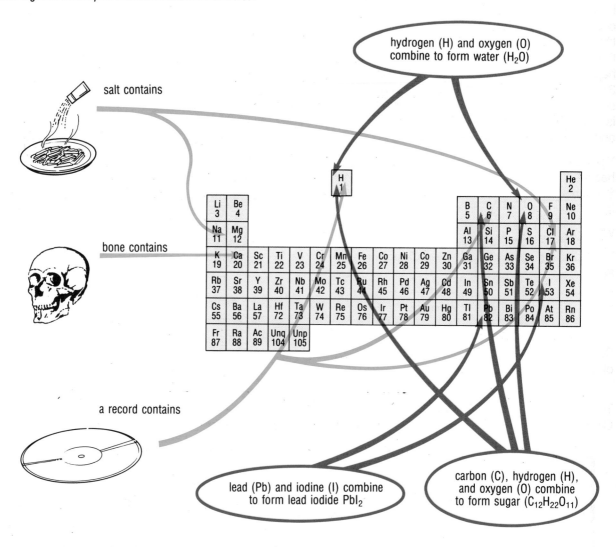

salt contains

bone contains

a record contains

hydrogen (H) and oxygen (O) combine to form water (H_2O)

lead (Pb) and iodine (I) combine to form lead iodide PbI_2

carbon (C), hydrogen (H), and oxygen (O) combine to form sugar ($C_{12}H_{22}O_{11}$)

4 How are symbols and formulas used to describe chemical reactions?

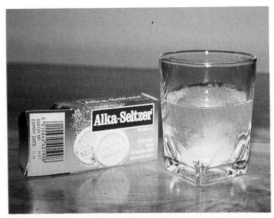

Figure 32 When an indigestion tablet (like Alka Seltzer) is added to water new substances are formed. ▶

◀ **Figure 31** When a match burns new substances are formed.

Both of the changes in figures 31 and 32 are called **chemical changes** because the starting chemicals, the **reactants**, have changed to different chemicals, the **products**. You can sometimes tell this has happened because of what you observe. In figure 31, heat and light are given out and a black substance which is different from the original match is left. In figure 32, a lot of bubbles of gas are formed and there was no gas there at the beginning.

You cannot always be sure that a chemical change has taken place by just observing what happens. Sometimes it is necessary to test what is left to show it is different from what you started off with.

When you work out what is happening, you can describe the change by writing a word equation.

For example, if a piece of charcoal (carbon) is burnt in oxygen you can test the gas formed to show it is carbon dioxide. The word equation for this change is:

carbon + oxygen → carbon dioxide

Sometimes symbols and formulas are used instead of words:

$$C + O_2 \rightarrow CO_2$$

This equation tells you that carbon is an element. C represents one atom of the element. Oxygen is also an element. O represents an atom of it but the formula O_2 shows that these atoms go around in pairs. Carbon dioxide is a compound because it contains two elements combined together and the formula CO_2 shows that they are combined in the ratio of one carbon atom to two oxygen atoms.

Overall the equation tells you that one C combines with one O_2 to form one CO_2.

The equation:

$$2Mg + O_2 \rightarrow 2MgO$$

tells you that 2 atoms of magnesium combine with one O_2 to form 2MgO.

When a chemical reaction occurs, no atoms are created and none are lost. They are just rearranged. That is why the 2 is placed in front of the Mg and in front of the MgO. The total number of atoms of each element on the left is equal to the total number on the right. For this reason it is called a **balanced equation**.

If you look in some chemistry books you will see lots of these equations. They are a quick way of describing chemical reactions and they provide more information than word equations. They are used by chemists throughout the world – they are the international language of chemists.

At this stage it is more important that you understand the information that the equation is telling you rather than be able to write equations yourself.

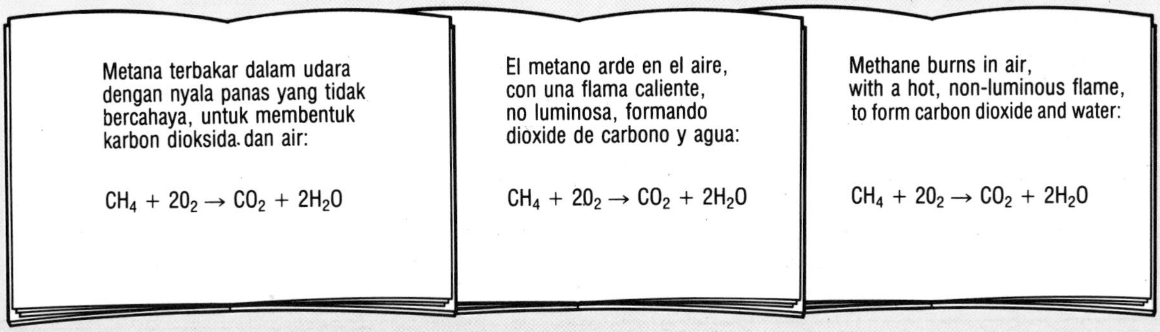

Metana terbakar dalam udara dengan nyala panas yang tidak bercahaya, untuk membentuk karbon dioksida dan air:

$$CH_4 + 2O_2 \rightarrow CO_2 + 2H_2O$$

El metano arde en el aire, con una flama caliente, no luminosa, formando dioxide de carbono y agua:

$$CH_4 + 2O_2 \rightarrow CO_2 + 2H_2O$$

Methane burns in air, with a hot, non-luminous flame, to form carbon dioxide and water:

$$CH_4 + 2O_2 \rightarrow CO_2 + 2H_2O$$

▲ **Figure 33** *The same equation is understood in any language.*

Things to try out

1 Baking powder, salt, talcum powder and sugar are four white solids. The following tests can be safely used on these substances at home. Use the tests to devise a key which could be used to distinguish between them.

Test 1 Shake a small quantity with water to see if it dissolves.

Test 2 If the substance dissolves, make a more concentrated solution. Test it with a piece of litmus paper.

Test 3 Add vinegar (which is an acid) to each.

Test 4 Line a baking tray with aluminium foil. Place a small heap of each substance at opposite corners of the tray – make sure you can remember which substance is which. Put the tray in an oven, set at 200°C and heat the substances for 15 minutes. Switch off the oven and leave it to cool. Then, using an oven-glove, carefully remove the tray and see if the heat has affected them differently.

Things to find out

2 Factories are often built in certain areas for good reason. For example:

In the north-east of England there are large underground deposits of calcium sulphate. This substance was used in chemical factories in nearby Billingham.

There are chemical factories at Runcorn which use the salt deposits found in that area.

There is a large oil refinery at Fawley which has a good deep sea access for tankers.

There is an aluminium extraction plant at Lochaber in Scotland which has a cheap source of hydroelectric power.

Trace an outline map of the British Isles. Use an atlas to find out where these factories are situated and then mark them on the map. Find out what the main products of these factories are and mark them on the map.

3 One of the Looking At sections of this chapter tells you something about John Dalton's contribution to our ideas on atoms. Marie Curie and Ernest Rutherford are two other famous scientists who each made important additions to

this theory. Use other books to find out about them and write one or two paragraphs on each of them pointing out how you think their work added to Dalton's theory.

Things to write about

4 Imagine you live in an area of high unemployment. A small chemical factory in the area wants to expand and produce a greater range of products. This will involve either more tankers travelling on the roads in the area or a new rail line being built to connect the factory to the main line. Write two letters to the local paper – one from a resident who lives near the road on which the tankers would travel and the other from a person through whose land the rail link would be built.

Making decisions

5 Use the Hazchem code to decide what course of action firemen should take if they arrive at an accident involving a tanker which carries this sign (figure 34):

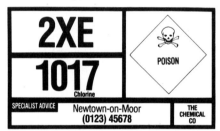

Figure 34

6 In some cases the reasons for originally building a chemical plant at a particular place are no longer important. Some of the factors which may have been important are:

close to raw materials,
good road, rail and sea links,
cheap energy source available,
close to customers,
reasonable distance from residential areas,
suitable workforce available,
a need to create jobs.

Select *two* of these factors which you think could, after a number of years, become less important. For each of them describe under what sort of circumstances this could happen.

Points to discuss

7 On your own, try to think of one object used at home which does not use a product of the chemical industry in its manufacture. Then discuss your ideas with each other and if you still think you have thought of an object that does not need the chemical industry, discuss it with your teacher.

8 The formula of vitamin C is $C_6H_8O_6$. What information does this formula tell you about the compound? Vitamin C, whether made artificially from oil or extracted from orange juice, still has the same formula. Think about this and discuss whether or not it is better to buy a drink with added artificial vitamin C.

Questions to answer

Questions 9-13

A SO_2 B C_2H_4O C $CaCO_3$
D $C_2H_4O_2$ E Cl_2

From the formulas A to E given above, choose the one which:

9 contains the element sulphur,

10 contains the element calcium,

11 contains only one element,

12 represents the greatest number of atoms,

13 contains the elements carbon, oxygen and hydrogen in the ratio of 2 : 1 : 4.

Questions 14–18

A $CH_4 + 2O_2 \rightarrow CO_2 + 2H_2O$
B $Mg + Cl_2 \rightarrow MgCl_2$
C $H_2SO_4 + Fe \rightarrow FeSO_4 + H_2$
D $H_2 + O_2 \rightarrow H_2O$
E $Fe + CuCl_2 \rightarrow FeCl_2 + Cu$

From the equations A to E given above, choose the one which:

14 represents a reaction which produces hydrogen,

15 contains the greatest number of different elements,

16 represents a reaction between a metallic element and a non-metallic element to form a compound,

17 represents the burning of a fuel to form carbon dioxide and water,

18 is *not* a balanced equation.

19 Use the properties which are listed below to construct a key which could be used to distinguish between substances A, B, C, D, E and F.

A **vegetable oil**
liquid, does not mix with water, flammable
B **salt**
white solid, soluble in water, solution conducts electricity and is neutral to litmus
C **lime**
white solid, reacts with water, solution turns litmus blue and conducts electricity
D **sugar**
white solid, dissolves in water, solution does not conduct electricity and is neutral to litmus
E **vinegar**
liquid, not flammable, mixes with water, solution turns litmus red
F **alcohol**
liquid, flammable, mixes with water, solution has no effect on litmus

20 Ammonia (NH_3) is made by reacting nitrogen (N_2) with hydrogen (H_2). Some ammonia is immediately reacted with another chemical, sulphuric acid (H_2SO_4) to form a fertilizer, ammonium sulphate [$(NH_4)_2SO_4$].

Some ammonia is also transported to another chemical factory where it is converted into nitric acid (HNO_3).

a) Draw a table using the headings given below and complete the table for all of the chemicals mentioned above.

Name of substance	Formula	Number of each type of atom	Is it an element or a compound?

b) Ammonia is a gas at room temperature, but it is easily changed to a liquid when it is cooled and compressed.
   (i) Using the idea of particles, explain the liquifaction of ammonia.
   (ii) Explain why it is better to transport ammonia in a road tanker as a liquid rather than as a gas.
   (iii) A dilute solution of ammonia can be used as a household cleaner. Explain what the difference is between liquid ammonia and a solution of ammonia.

c) In one part of the country, ammonia is transported by pipeline from one factory to another. Explain what factors are likely to influence the decision as to whether to transport ammonia by road tanker or by pipeline.

Introducing plastics

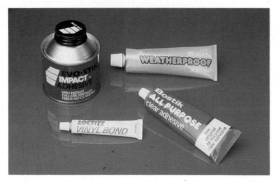

Figure 1 *What are the properties needed to make glues like these?*

The articles in figures 1–3 have one thing in common – they are made from **plastics**. They are made from different plastics which have different properties. It is these properties that determine how the plastics are used.

Plastics have become a very important part of modern life. The first plastic was used in 1843 to make buttons. Today plastics are used throughout industry for a wide variety of purposes.

In this chapter you will find out about:

◆ the uses and properties of plastics,
◆ the manufacture of plastics,
◆ the chemical structure of plastics.

▲ **Figure 2** *What are the properties of the plastics used to make these objects?*

Figure 3 *What are the properties of the plastic used to make coils of rope like these?*

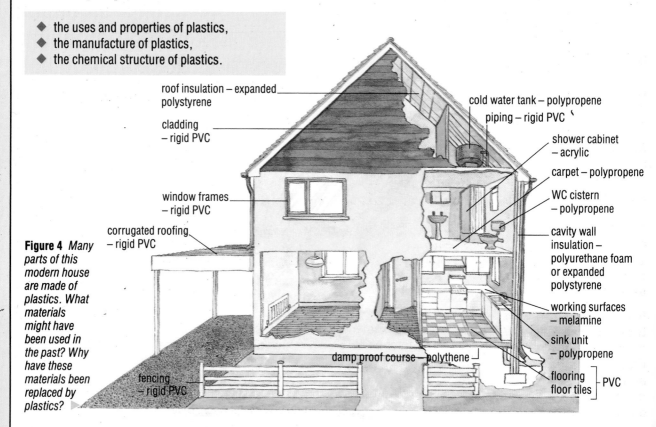

Figure 4 *Many parts of this modern house are made of plastics. What materials might have been used in the past? Why have these materials been replaced by plastics?*

roof insulation – expanded polystyrene

cladding – rigid PVC

window frames – rigid PVC

corrugated roofing – rigid PVC

fencing – rigid PVC

damp proof course – polythene

cold water tank – polypropene

piping – rigid PVC

shower cabinet – acrylic

carpet – polypropene

WC cistern – polypropene

cavity wall insulation – polyurethane foam or expanded polystyrene

working surfaces – melamine

sink unit – polypropene

flooring floor tiles – PVC

1 Plastics from crude oil

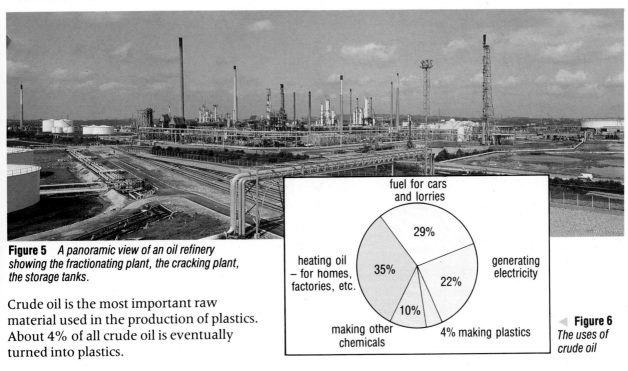

Figure 5 *A panoramic view of an oil refinery showing the fractionating plant, the cracking plant, the storage tanks.*

Crude oil is the most important raw material used in the production of plastics. About 4% of all crude oil is eventually turned into plastics.

Figure 6
The uses of crude oil

Separating crude oil – fractional distillation

Crude oil is a liquid containing many hundreds of different substances. Gases and solids are dissolved in the liquid.

Most of the substances in crude oil are **hydrocarbons**, compounds made up of carbon and hydrogen. It is very difficult to separate these hydrocarbons into *pure* compounds. Fortunately it is not often necessary to obtain absolutely pure compounds so they are only separated into groups with similar numbers of carbon atoms. This is done by fractional distillation using a fractionating column.

Inside an industrial fractionating column the crude oil is heated by a furnace where it evaporates. The vapours produced pass into the lower part of the fractionating column, as shown in figure 7. The vapours then pass up through horizontal trays with raised openings in them.

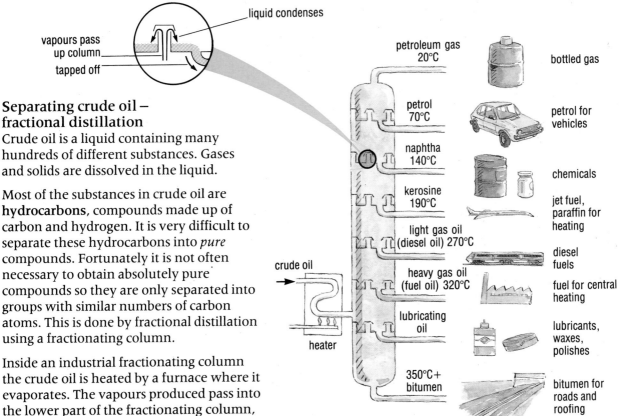

Figure 7 *The temperatures and products at different heights in an industrial fractionating column*

As the vapours rise up the column they cool and condense on to the trays where they are collected or **tapped off**. Different vapours condense at different temperatures and therefore at different heights in the tower. The liquids which are tapped off are called **fractions**. Each fraction contains hydrocarbon molecules which have a similar number of carbon atoms. Liquids like naphtha, which boil at low temperatures, do not condense until they get to the cooler parts near the top of the tower. Liquids like heavy gas oil (used in central heating) which boil at high temperatures, condense lower down in the tower where the temperature is higher.

Figure 8 *A fractionating column* ▷

Table 1 *The constituents and uses of the different fractions from crude oil*

Fraction	Boiling range	Number of carbon atoms in the hydrocarbons in the fraction	Uses
Petroleum gas	−160 to 40°C	1 – 4	Bottled gas, LPG, GAZ, plastics, chemicals
Petrol (gasoline)	40 to 140°C	5 – 10	Fuel for vehicles, chemicals
Naphtha	140 to 180°C	8 – 12	Chemicals, plastics
Kerosine (paraffin)	180 to 250°C	10 – 16	Jet fuel, chemicals
Light gas oil	250 to 300°C	14 – 20	Diesel fuel, chemicals
Heavy gas oil	300 to 350°C	20 – 30	Fuel for ships, factories and central heating
Lubricating oils, bitumen	above 350°C	more than 25	Lubricants, waxes, polishes, road surfaces, roofing

Plastics from oil fractions

The most important single substance for making plastics is ethene (figure 9). Ethene has only two carbon atoms. Chemists have found ways of obtaining ethene from the heavier fractions like naphtha and gas oil. These methods involve a process called **cracking**. During cracking larger molecules are broken down into two or more smaller molecules. Alkanes such as decane ($C_{10}H_{22}$), which are in the naphtha fraction, are broken down into smaller alkanes like octane (C_8H_{18}) and alkenes, such as ethene (C_2H_4).

decane   →   octane   +   ethene

$$C_{10}H_{22} \rightarrow C_8H_{18} + C_2H_4$$

Ethene and the other alkenes are used for plastics. The remaining 'cracked material' can be added to the petrol fraction or it can be further processed. Some of the important plastics and other products from ethene are shown in figure 9.

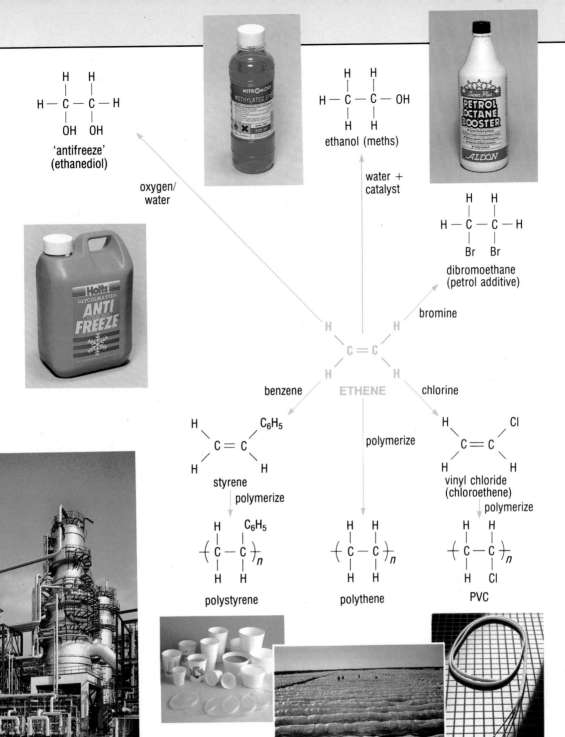

Figure 9 *Important plastics and other products obtained from ethene*

Figure 10 *A catalytic cracking plant at an oil refinery.*

How are molecules cracked?

Unlike distillation, cracking is a chemical process. It involves breaking a strong bond between two carbon atoms. This requires a high temperature. Naphtha is cracked by passing its vapour through very hot tubes (heated to about 800°C) where it is mixed with steam.

1 What is the difference between simple distillation and fractional distillation?
2 Why is crude oil fractionally distilled?
3 Why are the naphtha and gas oil fractions cracked?
4 Write an equation for the cracking of nonane (C_9H_{20}) to form ethene and an alkane.
5 What is the main difference between cracking and distillation?

2 Plastic waste

Plastic waste causes a lot of problems. Unsightly plastic rubbish like polythene bags, drinks containers and toffee wrappers, litter our streets. In the countryside cattle have choked after swallowing plastic bags. The problems are made worse because most plastic articles are not **biodegradeable**. This means that they do not decompose, like paper and wood, under the action of weather and bacteria. So plastic rubbish lies around for years and years (figure 11).

◄ Figure 11 Most of this waste is not biodegradable and should be properly disposed of - it won't disappear, it has to be removed and properly treated.

How do we deal with plastic waste?

Figures 12 and 13 show two of the ways in which we deal with most of our plastic waste at present – dumping it in holes in the ground and combustion.

Figure 12 A landfill site ►

Figure 13 An incineration plant

Recycling plastics, like recycling glass, sounds sensible. So why don't we treat all waste plastics like this? Some new research may help us to start recycling plastics. This process is described in the next section.

Plastics manufacturers are working on producing biodegradable plastics. Some biodegradable plastic shopping bags are already available.

1 Some people think that plastic waste is too valuable to dump in rubbish tips. What do you think should be done with it and why?

2 When plastics are burnt, acidic gases containing hydrogen chloride are sometimes produced. Which plastic do you think burns to produce hydrogen chloride?

3 Ford Motors at Dagenham produce about 300 tonnes of plastic waste each week. This is used in an incinerator to provide heating for the factory.
   a) Use table 2 to decide how many tonnes of heating oil the plastic waste will save. (Assume the plastic waste is all polythene and polystyrene.)
   b) What is the saving in cost? (Assume that 1 kg of heating oil costs £1.50.)

Table 2 The heat produced by burning 1 kg of different materials.

Material	Heat produced/kJ per kg
Polythene	
Polystyrene	45 000
Heating oil	
Fats	37 000
PVC	19 000
Paper	17 000
Wood	16 000

In many cities and towns, there are bottle-banks where you can put used glass bottles into large containers (figure 14). The bottles are then collected and melted down. The melt is used to remake bottles and other glass containers. There are no plastics-banks like these bottle-banks. Why is this?

All clear glass is made from the *same* chemical compounds. When melted down, the old glass produces a material which is easily re-used. The only real problem is in collecting the used glass bottles.

Unlike glass, plastic waste is a mixture of many *different* chemical compounds. Some plastics contain only carbon and hydrogen (polythene, polypropene, polystyrene), while others also contain chlorine (PVC) and oxygen and nitrogen (nylon). At present it is too expensive to separate all these different plastics for recycling, even though the idea of re-using plastics is very attractive.

Figure 14   *A bottle-bank where coloured glass is collected in separate containers.*

If plastic is heated in air, it burns producing mainly carbon dioxide and water. However, if plastic is heated in the *absence* of air, at about 700°C, the molecules break down to form smaller molecules. This method of cracking the large molecules is called **pyrolysis**.

Experiments to see whether it is possible to re-use plastic waste by using pyrolysis have produced encouraging results. A mixture of polythene, polypropene and polystyrene, for example, when cracked using pyrolysis gives gaseous hydrocarbons, such as methane, ethene and propene, and liquid hydrocarbons, such as benzene.

The gases, ethene and propene, can be separated by distillation and used to make chemicals, including plastics. The liquids can be used directly as fuels or as starting materials for other products.

Experimental pyrolysis plants such as the one in figure 15 can recycle 8000 tonnes of plastic a year. This is a small amount compared with the total worldwide production of plastics each year of over 75 million tonnes.

At present, the process is uneconomic. It is still cheaper to make the gases and liquids required to produce plastics from crude oil.

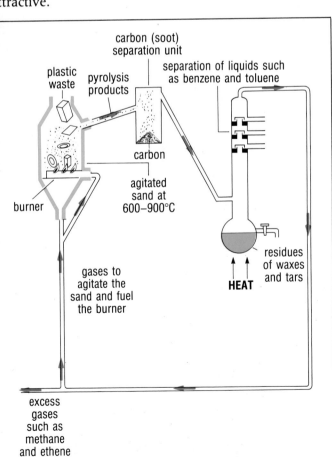

Figure 15   *An experimental pyrolysis plant*

However, as oil gets scarcer and more expensive, attempts to re-use plastics will become more important. This is why recycling research is now being carried out.

It is important to recycle plastics because

♦ *it would save **raw materials*** – many plastics are made from oil,
♦ *it might save **money*** – when oil becomes even more expensive,
♦ *it would save **the environment*** – recycling will reduce the amount of plastic litter.

1 How does the pyrolysis plant in figure 15 work after the plastic is inserted?
2 What are the main problems in recycling plastic waste?
3 Suppose the local authorities in your town/city want to collect plastic waste for recycling. What advice would you give to ensure that the recycling is done as cheaply as possible?

4 Designing a plastic

Plastics are used for many different purposes. A very rigid plastic is needed for containers of household chemicals, such as bleach and detergents. Softer plastics are used to make supermarket shopping bags, food bags and food wrapping sheets. Both types of plastics can be made from polythene.

Polythene is formed by the polymerisation of ethene. Polythene molecules consist of long chains of carbon atoms with hydrogen atoms attached. The chains contain up to 50 000 carbon atoms.

$$-C-C-C-C-C-C-C-C-C-C-C-C-$$

After polythene has been made, it sets, on cooling, to produce a solid. In this solid, the chains line up closely with each other.

Because of the way it sets, the plastic has quite a high softening point and a high density and can be used to make rigid containers. To produce a softer plastic small amounts of another alkene are added to ethene *before* polymerisation.

The structure of the added alkene is

$$CH_3 - C = CH_2$$
$$\quad\quad | \quad CH_3$$

This alkene contains 4 carbon atoms but they are not in a straight chain. When ethene polymerises, the second alkene is incorporated into the polymer:

The **CH₃** groups occur at irregular points along the chain. Because of this, polymer molecules cannot get as close to each other when the molten polymer sets to form a solid.

This polymer therefore has a lower softening point than polythene. It can also be pressed into a film more easily.

Find the name of the alkene with the formula:

$$CH_3 - C = CH_2$$
$$\quad\quad | \quad CH_3$$

Would you expect the new polymer to have a higher or a lower density than polythene? Explain your answer.

1 Several plastics and many other materials obtained from coal, crude oil and natural gas are hydrocarbons. Hydrocarbons are compounds containing carbon and hydrogen only.

2 The simplest series of hydrocarbons are alkanes. The first four members of the alkane series are:

	methane	ethane	propane	butane
Molecular formula	CH_4	C_2H_6	C_3H_8	C_4H_{10}
Structural formula	H | H–C–H | H	H H | | H–C–C–H | | H H	H H H | | | H–C–C–C–H | | | H H H	H H H H | | | | H–C–C–C–C–H | | | | H H H H
Molecular model				

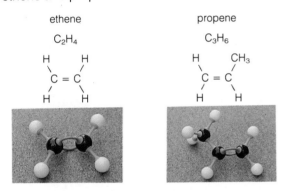

Another series of hydrocarbons are called alkenes. All alkenes contain a $>C = C<$ bond and their names end in −ene. The simplest alkenes are ethene and propene.

ethene	propene
C_2H_4	C_3H_6
H H \\ / C = C / \\ H H	H CH_3 \\ / C = C / \\ H H

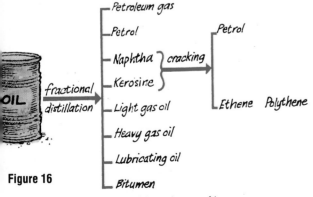

Alkenes can be identified because they decolorize bromine water. They are unsaturated hydrocarbons.

3 Polythene and several other plastics are made from ethene which is obtained by cracking the heavier fractions from crude oil.

Figure 16

— Petroleum gas
— Petrol
— Naphtha ⎫ cracking → Petrol
— Kerosine ⎭
— Light gas oil → Ethene — Polythene
— Heavy gas oil
— Lubricating oil
— Bitumen

OIL — *fractional distillation* →

Cracking is the breaking down of large hydrocarbon molecules into smaller ones by the action of heat.

4 There are two different kinds of plastics:
   a) **Thermoplastic** polymers like polythene, PVC and nylon which soften (become plastic) on heating. These consist of long, thin chains without crosslinks, as shown in figure 17a. They have low melting points, melt without decomposing and stretch easily.

a)  b)

Figure 17

   b) **Thermosetting** polymers like Bakelite, urea-formaldehyde and melamine which do *not* soften on heating. These plastics consist of cross-linked chains, as shown in figure 17b. They have high melting points, decompose on heating and are usually rigid

5 Different plastics have different physical and chemical properties. They can be identified by properties such as:
   a) density — Does it float on water?
   b) hardness — Can it be marked by an HB pencil?
   c) combustion — Does it burn with a smokey flame? Is ash left?

6 Plastics have a great variety of uses. These uses of plastics are related to their properties.

7 Plastics (like polythene, perspex and PVC) are polymers. The plastics are made by polymerisation which involves joining together small molecules, called monomers, to form a long chain molecule, called a polymer.

(M) + (M) + (M) + (M)   –(M)–(M)–(M)–(M)–

monomer molecules   part of the polymer molecule

Some polymers are made from two different monomers.

8 Polythene is the most used plastic. It is made by polymerisation of ethene.

ethene monomers → part of polythene polymer

The monomers for perspex, PVC, polystyrene and polypropene all contain double bonds between carbon atoms like ethene. These double bonds enable the monomers to polymerise.

9 Most plastics do not decay. They are non-biodegradeable. This means that plastic waste can cause a pollution problem.

1 What are plastics?

There are about 20 important types of plastic but these are produced commercially under more than 5000 trade names. Because of this, it is sometimes difficult to recognize plastics from their trade names.

Some of the most important plastics and their monomers are shown in table 3.

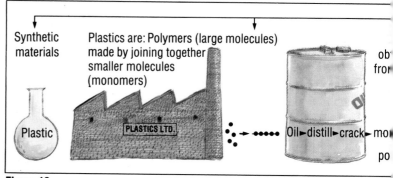

Synthetic materials

Plastics are: Polymers (large molecules) made by joining together smaller molecules (monomers)

ob
fro

Plastic

PLASTICS LTD.

Oil►distill►crack► mo

po

Figure 18

Table 3 *Some important plastics and their uses*

Plastic	Monomer(s)	Uses
Polythene	ethene $H_2C=CH_2$	food bags, shopping bags, insulation for electrical wiring
Polypropene (polypropylene)	propene $CH_3CH=CH_2$	car battery housings, piping, bottle crates, carpets
PVC	vinyl chloride (chloroethene) $CHCl=CH_2$	records, artificial leather, water pipes
Polystyrene	styrene $C_6H_5CH=CH_2$	packaging, foamed material
Perspex (acrylic)	methylmethacrylate $CH_2=C(CH_3)-C(=O)-O-CH_3$	safety glass, reflectors, traffic signs, contact lenses, false teeth
PTFE	tetrafluoroethene $F_2C=CF_2$	non-stick surfaces (pans, skis), gaskets
Nylon (polyamide)	diaminohexane $H_2NCH_2CH_2CH_2CH_2CH_2CH_2NH_2$ and hexanedioyldichloride $Cl-C(=O)CH_2CH_2CH_2CH_2C(=O)-Cl$	rope, clothing (tights), carpets
Terylene	ethanediol $HO-CH_2CH_2-OH$ and terephthalic acid $HO-C(=O)-C_6H_4-C(=O)-OH$	clothing (trousers, shirts), sails, ropes

2 Why are plastics important?

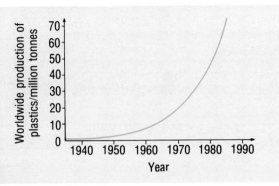

Figure 19

The worldwide production of plastics has risen dramatically over the last 50 years and now 75 million tonnes a year are produced (figure 19).

About one third of all plastics are used in packaging. A small car has a mass of about 800 kg and about 100 kg of this is made of plastics. The building industry uses the largest amount of plastics produced each year.

Different plastics have different properties. In fact, one of the great advantages of plastics is that they can be *made* to have particular properties. Plastics are chosen for their different uses because they have a suitable combination of properties. Plastics can be:

Cheap   Easily moulded
Strong   Resistant to chemicals
Light   Flexible
Clear   Dyed

Figure 20 shows how the uses of some plastics are related to their properties.

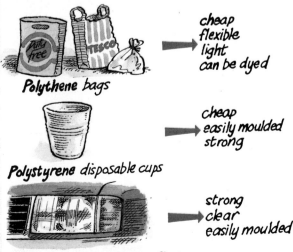

cheap
flexible
light
can be dyed

Polythene bags

cheap
easily moulded
strong

Polystyrene disposable cups

strong
clear
easily moulded

Perspex (acrylic) car lights reflector

Figure 20

3 How are plastics made?

Plastics are made by polymerisation.
There are two kinds of polymerisation:

◆ addition polymerisation and
◆ condensation polymerisation.

Addition polymerisation

Addition polymerisation involves monomers like ethene, vinyl chloride (chloroethene) and styrene which contain double bonds between carbon atoms. During polymerisation, the double bonds between pairs of carbon atoms 'open up' and the carbon atoms of separate ethene molecules join together to form a molecule of polythene.

$$... + \begin{matrix} H \\ \diagdown \\ \diagup \\ H \end{matrix} C = C \begin{matrix} H \\ \diagup \\ \diagdown \\ H \end{matrix} + \begin{matrix} H \\ \diagdown \\ \diagup \\ H \end{matrix} C = C \begin{matrix} H \\ \diagup \\ \diagdown \\ H \end{matrix} + \begin{matrix} H \\ \diagdown \\ \diagup \\ H \end{matrix} C = C \begin{matrix} H \\ \diagup \\ \diagdown \\ H \end{matrix} + ...$$

ethene molecules
↓

$$\begin{matrix} H & H & H & H & H & H \\ | & | & | & | & | & | \\ \sim C - C - C - C - C - C - \\ | & | & | & | & | & | \\ H & H & H & H & H & H \end{matrix}$$

part of a molecule of polythene

In this way, a very long chain of carbon atoms is formed to give polythene. The chain may contain more than 50 000 carbon atoms. If n molecules of ethene combine, the process can be written as

$$n \left(\begin{matrix} H \\ \diagdown \\ \diagup \\ H \end{matrix} C = C \begin{matrix} H \\ \diagup \\ \diagdown \\ H \end{matrix} \right) \rightarrow \left(\begin{matrix} H & H \\ | & | \\ - C - C - \\ | & | \\ H & H \end{matrix} \right)_n$$

Other polymers are also made from monomers containing carbon-carbon double bonds.

$$n \left(\begin{matrix} H \\ \diagdown \\ \diagup \\ H \end{matrix} C = C \begin{matrix} Cl \\ \diagup \\ \diagdown \\ H \end{matrix} \right) \rightarrow \left(\begin{matrix} H & Cl \\ | & | \\ - C - C - \\ | & | \\ H & H \end{matrix} \right)_n$$

vinyl chloride   PVC
(chloroethene)

$$n \left(\begin{matrix} H \\ \diagdown \\ \diagup \\ H \end{matrix} C = C \begin{matrix} C_6H_5 \\ \diagup \\ \diagdown \\ H \end{matrix} \right) \rightarrow \left(\begin{matrix} H & C_6H_5 \\ | & | \\ - C - C - \\ | & | \\ H & H \end{matrix} \right)_n$$

styrene   polystyrene

Condensation polymerisation

The second way to make a polymer is by condensation polymerisation. In this process, two compounds with *reactive* atoms at the ends of their molecules react together. Nylon, terylene and urea-formaldehyde resin are condensation polymers.

Nylon is made from

$$
\begin{array}{c}
H \\ \backslash \\ N-C-C-C-C-C-C-N \\ / \\ H
\end{array}
$$

1, 6-diaminohexane

which can be represented as   H—◯—H

and

$$
O=C-C-C-C-C-C=O \\
Cl \qquad Cl
$$

hexanedioyldichloride

which can be represented as   Cl—▭—Cl

When a molecule of 1, 6-diaminohexane reacts with a molecule of hexanedioyldichloride the two join together and hydrogen chloride is also produced.

H—◯—H + Cl—▭—Cl ⟶ H—◯—▭—Cl + HCl

This process can occur again

H—◯—▭—Cl + H—◯—H ⟶ H—◯—▭—◯—Cl + HCl

and again and again, forming a long polymer molecule.

In the examples so far the chains only extend in one direction – lengthways. Sometimes, the long polymer molecules can form bonds with each other at places along their length. This produces long chains which are cross-linked giving them a more rigid three-dimensional structure.

4   How are the properties of plastics related to their structure?

There are two different kinds of plastics. One kind contains **long, thin molecules which form tangled chains**. Examples of these plastics are polythene, polystyrene, acrylics, polypropene, PVC and nylon.

In these plastics, there are strong forces between the atoms along the chain but much weaker forces *between the tangled chains*.

This structure gives these plastics very important properties. Plastics containing long, thin, tangled chains have

◆ weak forces between chains,
◆ chains which move over each other easily.

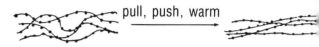
pull, push, warm

These plastics:

stretch easily
flex easily
soften on warming
melt at low temperatures
melt without decomposing

They are called **thermoplastics**. These plastics can be shaped by warming and then moulding, pressing or extruding.

The second kind of plastics contain **large cross-linked molecules**,

Examples of these are Bakelite, urea-formaldehyde resin, phenol-formaldehyde resin, melamine-formaldehyde resin and epoxy glues.

In these cross-linked plastics there are strong forces holding the atoms in chains but there are also strong forces between the chains.

This structure gives these plastics very different properties.

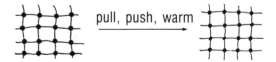
pull, push, warm

Plastics containing large, cross-linked molecules have

◆ strong forces holding the atoms in the chains and also strong forces between the chains
◆ chains which cannot move relative to one another on pulling, pushing or warming.

These particular plastics:

> are rigid and hard
> do not flex
> burn or char before melting

They are called **thermosets**.

Figure 21 *A selection of plastics – find out which are thermoplastics and which are thermosets.*

Things to do

Things to try out

1 *Comparing the conduction of heat by plastics and metals*

   Design an experiment that you could do at home to see how well plastics conduct heat compared to metals.
   a) Describe the experiment you would do.
   b) What would you do to get a fair comparison?
   c) What results would you expect?

2 *Using models to understand cracking*
   a) Use a molecular model kit to make a molecule of hexane (C_6H_{14}). This contains a straight chain of six carbon atoms with hydrogen atoms attached.
   b) Imagine that your hexane molecule undergoes cracking to form ethene. Break your hexane model to represent this process. What other product is formed besides ethene?
   c) Make a model of hexane again. In what other ways can the hexane molecule be cracked? What products are produced in these cases?

Things to find out

3 a) Which parts of the world does crude oil come from?
   b) How did it form below ground in the first place?

4 a) What does the term 'organic substance' mean?
   b) Why is this term used to describe plastics?

5 Plastics are processed and shaped by extruding, injection moulding, calendering, foaming, laminating and thermoforming. Find out more about one or two of these processes.

Points to discuss

6 Some people think that oil is our most important raw material. What do you think? Are there any raw materials more important than oil?

7 What will life be like when the supply of crude oil runs out?

8 Some people say that crude oil is much too valuable to burn. Giving your reasons, decide whether you agree or disagree with them.

9 If the molecules in a polymer are linked by a *small number* of bonds to form a loose network, a rubber-like material is formed. This is called an **elastomer**.
   Elastomers are more elastic than thermoplastics and thermosets. Elastomers return to their original shape after stretching. Thermoplastics do not return to their original shape after stretching. Thermosets are hard and rigid.

a) How do you explain these properties of elastomers in terms of their structure?
b) Why are elastomers more elastic than thermoplastics?
c) Why are thermosets hard and rigid?

10 In some cars, about 30% of the components used are made from plastics. Why do car manufacturers use so much plastic?

11 'We should prohibit the use of disposable plastic containers and only permit the returnable type.' What do you think? In your discussion, think about the advantages and disadvantages of:
a) disposable plastic containers and
b) returnable containers.

12 Andrew prefers to buy lemonade in a glass bottle. Carol prefers to buy lemonade in plastic bottles. Which do you prefer? Give reasons for your choice.

Making decisions

13 Nowadays, metal articles are often coated by dipping them in molten plastic rather than enamelling or electroplating them. Four plastics which might be used in this process are PTFE, polythene, perspex and urea-formaldehyde resin. Some properties of these four plastics are shown in table 4.

Which plastic would be best for:
a) coating the handles of small screwdrivers,
b) coating electrical wiring,
c) coating the inside surface of a frying pan?

Explain your choice of plastic in each case.

14 Look around the room in which you are working.
a) Identify four articles made of plastic.

b) Why is each article made of that particular plastic?
c) What material do you think each article would be made of if plastics did not exist?
d) What are the advantages of making the articles from plastic?

Questions to answer

15 The apparatus in figure 22 can be used to break down plastic X which is like polythene.

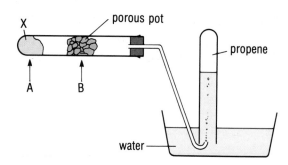

Figure 22

a) Why is the heat applied first at B and then at A?
b) Why will the first test tube of gas collected contain air as well as propene?
c) Draw the structural formula of propene.
d) What is the chemical name of X?
e) What would you see when you add bromine water, drop by drop, to a test tube containing propene and shake the mixture?
f) What do you think will happen when X is shaken with bromine water?
g) Why do propene and X react differently with bromine water?

Table 4 *Some properties of four plastics*

Plastic	Type of plastic	Strength relative to low density polythene	Flexibility	Max temp for use/°C	Resistance to:		Cost /£ per kg
					dilute acids	oily substances	
PTFE	thermoplastic	30	fairly flexible	250	excellent	excellent	20.0
Polythene (low density)	thermoplastic	1	very floppy	70	good	good	0.7
Perspex	thermoplastic	9	stiff	90	good	good	1.6
Urea-formaldehyde resin	thermoset	9	very stiff	75	poor	good	0.9

Table 5 *Some data about plastics*

Plastic	Touch with a red-hot nail	Heat with a burning splint	Smell while burning
Nylon	melts	burns without smoke, ash is left	smell of burning hair
Polystyrene	melts	burns with a smokey flame, producing black smuts	no distinct smell
PTFE	melts	does not burn	no distinct smell
PVC	melts	burns with a smokey flame	unpleasant sharp smell
Urea-formaldehyde resin	chars, does not melt	swells and cracks	fishy smell

16 Table 5 gives some properties of five plastics. Use the results in the table to make a key to identify each of the five plastics.

17 a) In what order or pattern are the hydrocarbons listed in table 6?

Table 6 *Boiling points of some hydrocarbons*

Hydrocarbon	Boiling point /°C
Butane	1
Decane	174
Ethane	−88
Hexane	69
Octane	126

b) Suppose you wanted to see if there was a trend in the boiling points. How would you list the hydrocarbons?

c) Rewrite the data according to your answer in part (b).

d) Which hydrocarbons are gases at room temperature (20°C)?

e) Butane and propane (b.pt. −41°C) are used in steel containers as fuel by campers. Why is it dangerous to use butane during the winter?

18 The following list shows the uses to which plastics are put.

Packaging	36%
Building industry	21%
Electrical and electronic industry	10%
Transport industry	5%
Furniture industry	5%
Toys and leisure items	4%
Other uses	19%

Draw a pie chart to summarize these uses.

19 Table 7 gives some data concerning polythene and polystyrene.

a) Draw a section of the polymer containing 3 monomer units for i) polythene, ii) polystyrene.

b) Why do you think that the density of polystyrene is larger than that of polythene?

c) Why do you think that polythene stretches more easily than polystyrene?

d) Why do you think that polythene is soft but polystyrene is brittle?

Table 7 *Some data concerning polythene and polystyrene*

Polymer	Monomer	Density of polymer /g cm^{-3}	Strength/hardness of polymer
Polythene	ethene $\begin{array}{c} H \quad\quad H \\ \ \ \diagdown \quad \diagup \\ \ \ C = C \\ \ \ \diagup \quad \diagdown \\ H \quad\quad H \end{array}$	0.94	stretches easily, soft
Polystyrene	styrene $\begin{array}{c} H \quad\quad C_6H_5 \\ \ \ \diagdown \quad \diagup \\ \ \ C = C \\ \ \ \diagup \quad \diagdown \\ H \quad\quad H \end{array}$	1.06	does not stretch so easily, brittle

Introducing growing food

If you or your family have ever tried to grow vegetables you will know that slugs, caterpillars and weeds are just three of the problems you have to contend with! Sometimes the local wildlife eat more of your cabbages and lettuces than you do. If you live on a farm you will know even more about the havoc that can be caused by pests, bad weather and changing market prices.

Figure 1 *Ladybirds and other predators help the farmer by destroying some pests but most pests and weeds on farms are controlled by using agrochemicals.*

Figure 2 *Birds cause considerable damage to some crops and various methods are used to scare them away. What other methods do farmers use?*

Figure 3 *Spraying with agrochemicals helps to produce disease-free and undamaged crops. What are the advantages and disadvantages of the various methods of application?*

Figure 4 *Farmers, with the help of chemists and biologists, have been so successful in recent years in their battles with pests and diseases that there are now surpluses of some crops which have to be kept in stores like the one in the photograph, until required.*

Growing food is a hazardous business but it is absolutely vital for our survival. So not surprisingly scientists have devoted a great deal of time and money to developing ways of increasing crop production. Chemistry has had a considerable contribution to make. Chemists have developed chemicals which kill pests and increase the fertility of the soil. These chemicals are called **agrochemicals** and manufacturing them forms the basis of one of the largest industries in the world.

In this chapter you will see

- what is in soil which helps plants to grow healthily,
- what is in fertilizers, whether they are necessary and, if so, how they can be made more cheaply,
- how plants can be protected from insects, weeds and diseases,
- what possible side-effects there are when synthetic chemicals are used in agriculture and how these side-effects are kept to the minimum.

Using chemicals to save plants

Farmers and gardeners face severe competition in trying to raise good yields of healthy crops. Crops are attacked by a multitude of insects, slugs, snails, worms and other animals, which want the plants as their food. Bacteria and fungi may also feed on the plants causing them to become diseased. Weeds invade the growing area and compete with the main crop for the nutrients and water which the plants need.

To help the growers to fight back against all these forms of attack, chemicals called **pesticides** can be used (figure 5).

Farmers in Asia, Africa and South America can lose 40% of their crops to pests; in Europe the loss is about 25%. So you can see what a vital part pesticides have to play in helping to control these huge losses.

Controlling harmful insects

Many insects, such as bees, are essential to the life of plants, as they carry pollen from flower to flower. Others however, bite, suck and chew and cause so much loss or damage to plants and farm animals, that farmers must control them.

Some plants contain their own insecticide. For example, compounds called **pyrethrins**, are extracted from a type of chrysanthemum. Another insecticide is **derris** which is extracted from a tropical climbing plant.

These natural insecticides are deadly to many insects but do not poison humans or most animals, although they can be harmful to fish. They may be used safely. However, their effect does not last for long and farmers tend to use cheaper man-made (also called synthetic) insecticides.

DDT and beyond

The most famous synthetic insecticide is **DDT** (Dichloro Diphenyl Trichloroethane) (figure 6).

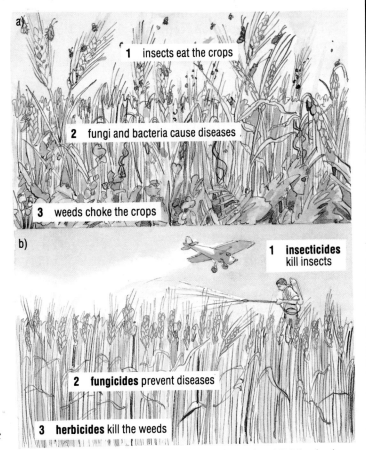

a)
1 insects eat the crops
2 fungi and bacteria cause diseases
3 weeds choke the crops

b)
1 **insecticides** kill insects
2 **fungicides** prevent diseases
3 **herbicides** kill the weeds

Figure 5 *a) A three pronged attack and b) three ways of fighting back.*

Figure 6 *A DDT molecule*

The story of DDT is an example of the problems humans face in using chemicals on a large scale in the environment.

DDT is a very effective insecticide. It was rapidly introduced during World War II to keep soldiers free from lice. After the war, DDT was used in vast amounts for killing disease-carrying insects, such as mosquitoes, and insect pests on plants. Many millions of human lives have been saved by the action of DDT and similar compounds.

Then disturbing information began to arrive. Fish were dying and the number of birds of prey, in particular, was getting smaller. DDT was also found to be building up in the fatty tissues of animals. The reason was not hard to find.

DDT is unreactive and eventually is washed into rivers, lakes and the sea. In the water there are billions of small organisms, called **phytoplankton** living near the surface. They need water, carbon dioxide, nutrients and sunlight, but will also absorb DDT.

Feeding on the phytoplankton are very small animals, the **zooplankton**, for example, water fleas. These in turn are eaten by small fish, and so on. The fish can be eaten by birds and humans. Once the DDT is taken in by the plankton (figure 7), fish, birds or humans, it is not removed as it dissolves much more readily in fats and oils in the animals bodies than in water. In figure 8 you will see that the water may contain 0.1 parts of DDT in 100 000 000 (1 billion) parts of water. This will mean that a large fish could contain 1 part of DDT in 1 000 000 parts of the fish. The DDT has been **concentrated** 1000 times.

DDT is a poison and there is great concern about its effects on birds, such as herring gulls, pelicans and ospreys, that live predominantly on fish. A particular problem is that the affected birds lay eggs with very thin shells, which break before the young birds hatch and so the next generation of the birds is lost.

Great public concern arose about the long-term effects of DDT and, in the early 1970s, DDT and similar compounds were withdrawn from use in this country. However, some poorer countries still use it because it is a very cheap insecticide to make. They argue that it is more important for them to kill insects which spread malaria and other diseases, than to worry about the long-term effect on fish and birds and ultimately humans.

The countries which banned DDT began to use more pesticides containing phosphorus – the organo-phosphorus pesticides. These break down more rapidly than DDT and do not concentrate through the food chain. They can be **systemic**; that is, they can be taken up inside a plant, so that the whole plant becomes poisonous to insects which eat it. The drawback with these organo-phosphorus insecticides is that they are very poisonous to humans and animals. Farmers must be very careful when they are spraying them on crops in the fields. They are much more poisonous than DDT.

However, unlike DDT, organo-phosphorus compounds break down in the soil and do not have the long-term dangers of DDT which were described above.

Figure 7 Plankton, like these, are food for large creatures in the food chain.

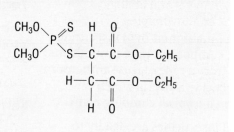

Figure 9 A malathion molecule. This is called an organophosphorus compound because it contains both carbon and phosphorus.

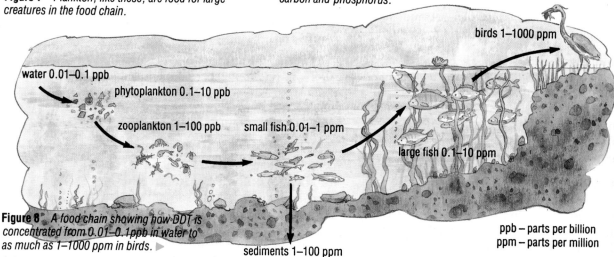

Figure 8 A food chain showing how DDT is concentrated from 0.01–0.1ppb in water to as much as 1–1000 ppm in birds. ▶

water 0.01–0.1 ppb
phytoplankton 0.1–10 ppb
zooplankton 1–100 ppb
small fish 0.01–1 ppm
large fish 0.1–10 ppm
birds 1–1000 ppm
sediments 1–100 ppm

ppb – parts per billion
ppm – parts per million

Looking for new insecticides

We have seen one danger in using insecticides – they may not only poison the insects they are used against but they sometimes poison other living things too. Another problem with insecticides is **insect resistance**. Most insects may be killed by the fairly low doses of an insecticide when it is first introduced. After some years of use, larger doses are needed and a few of the insects are still not affected. These unaffected insects breed rapidly as they face less competition. Soon the original size of the insect population is restored and *all* the insects are now resistant to the insecticide.

Chemical firms spend a great deal of research time and money looking for new insecticides, particularly as insects become resistant to the older ones. It is very difficult to find a chemical which is effective against an insect and yet completely safe for other animals. Thousands of new compounds are tested each year and it takes up to ten years for an insecticide to progress from discovery to marketing. Because of this time scale the manufacturers are only just keeping ahead of the insects' ability to become resistant.

Some recent chemical and biological developments are now helping to reduce the need for insecticides. These include:

◆ breeding plants whose taste the attacking insects will dislike,
◆ using bacteria and viruses to 'fight bugs with bugs'.

Our insect competitors may greatly outnumber us but we are fighting back! In the next section, we look at another method of insect control.

Beware of sex!

Animals and insects release compounds to which other members of the same family respond. The compounds act as signals, for example to attract the opposite sex for mating. These compounds are called **pheromones** and are now being used to help control insects.

An example is the use of a pheromone to control the gypsy moth which causes great damage to trees in the US. The female gypsy moth gives out a pheromone which attracts the male moths. Chemists have studied this

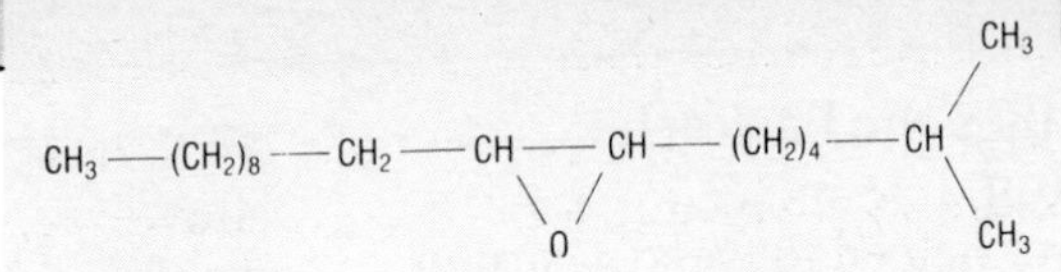

$$CH_3 - (CH_2)_8 - CH_2 - CH - CH - (CH_2)_4 - CH \begin{smallmatrix} CH_3 \\ \\ CH_3 \end{smallmatrix}$$
$$\underset{O}{\diagdown}$$

◀ **Figure 10**
The pheromone produced by the female gypsy moth

pheromone and have discovered its formula and structure (figure 10). A method of making the pheromone from simple substances – '**a synthesis**' – has been discovered.

One way of reducing the gypsy moth population (figure 11) is to put a tiny amount of the synthetic pheromone into a trap which also contains a sticky substance (figure 12). The male moth is attracted by the pheromone and gets stuck in the trap and dies.

Figure 11 ▼

Figure 12 ▲
A pheromone trap in an apple tree

Another way of controlling the moths is simply to monitor the number of moths trapped each day. When the number increases to a dangerous level, insecticides are used to reduce them. This technique allows insecticides to be used only when absolutely necessary.

Another pheromone is being used to help control a beetle which attacks spruce trees in the forests of Norway and Sweden. It is estimated that 4 billion beetles are captured this way each year.

Much research remains to be done into this method of insect control but it may be more widely used in the future.

---

1 What is your opinion about the use of insecticides by farmers and gardeners?
2 Which of the methods of insect control that have been discussed in this section do you think chemists and biologists should concentrate their research on and why?

Chemistry takes the backache out of weeding

If you have ever helped to weed a garden, you will know how tiring it is. (Imagine trying to weed a huge field of vegetables or wheat!) Worse still, a few weeks later, the weeds will be back and the area will look just as bad again. But it is not just the look of the area that make it important to get rid of weeds. The weeds will take away much of the food in the ground – the nutrients and water – that the crops need. Some weeds, like bindweed, physically choke the plants. To get rid of these unwanted weeds in cultivated ground chemists have developed a wide variety of weedkillers or **herbicides**.

Total herbicides kill all green vegetation. Sodium chlorate has been used for this purpose for many years. The best known of the total herbicides is **paraquat**. This is particularly useful because it kills vegetation on contact and then breaks down quickly in the soil and becomes harmless. It can be used to kill weeds before the main crop seedlings appear. Paraquat is extremely poisonous to animals, including humans, and must be used and stored very carefully.

Selective herbicides are active only against particular weeds, for example, nettles. One selective herbicide called **2,4,5-T** caused much concern when it was thought to be involved in birth defects in animals. As a result it was banned in many countries.

However, the birth defects have now been shown to be due to small quantities of an impurity in 2,4,5-T called dioxin. New methods of manufacturing 2,4,5-T remove the traces of dioxin and may allow it to be used again safely.

> When would a farmer be likely to use
> a) paraquat and b) 2,4,5-T?

Fighting fungus – save our spuds!

After protecting their crops from insects and weeds, farmers still cannot relax. Fungal diseases also attack many food plants and can spread through a crop very quickly (figure 14). There have been some disastrous examples.

Figure 13 The weeds in this cereal stubble (below) are destroyed (left) by a herbicide like paraquat before the field is cultivated for the next crop.

Potato 'blight' caused the Great Famine in Ireland in the years between 1846 and 1849. The poor people lived largely on potatoes and when their crops were destroyed by the blight a million people starved to death and many were forced to emigrate. The history of Ireland, and indeed the US, where many of the Irish went, was changed by the blight.

In the 1950s, 80% of the wheat crop in the US was wiped out by a fungus called 'wheat rust'.

Figure 14 If the fungus on this wheat is allowed to spread it will cause serious damage to the whole crop. By using a fungicide the farmer can prevent this.

These are very severe examples of the fungal attacks on crops which are going on all the time. It is estimated that about 60% of the food in the world is destroyed by fungal diseases before it can be used: about 30% as it grows and about 30% in storage.

Fungal diseases are now being controlled by breeding plants which can resist the fungus (figure 15). This takes time and meanwhile farmers have to use chemicals, called **fungicides**. Potato blight was eventually controlled by spraying potato plants with 'Bordeaux Mixture' (a mixture of copper(II) sulphate and calcium hydroxide in water). However fungi, like insects with insecticides, can become resistant to fungicides. Potato blight is at present only just under control, despite the development of new fungicides.

Figure 15 *Two new wheat varieties being tested in field trials. Disease resistance is an important characteristic when new varieties are introduced.*

2 Farming without chemicals?

Some health-food shops claim that the food they sell has been grown 'organically – without the use of chemicals'. As all plants use chemicals from the soil to grow, this claim is not accurate. However, what the shops and organic farmers mean is that they do not use any **synthetic chemicals** as either fertilizers or pesticides as they consider that these synthetic chemicals are harmful.

You may also have seen claims that organically-grown food is more nutritious than food grown with the help of synthetic fertilizers. However, scientific tests can not detect any difference between them. Natural organic fertilizers – compost and manure – are broken down to the same ions used by plants as the synthetic inorganic chemicals. There is no real advantage, in terms of food nutrition, for organic farming. However, as there are small amounts of pesticides on the surface of the foods which are not grown organically, it is important to wash all fruit and vegetables before eating them.

The yields of most organic farms are lower than those of traditional farms. This means that the food they produce is more expensive. The amount of both fertile land suitable for growing organic crops and of natural fertilizer is limited. Also most

Figure 16 *How do organic farms differ from traditional ones?*

agricultural produce is consumed in large cities but it is difficult to return the waste material (sewage, etc) produced in the cities to agricultural areas where it could be used to fertilize the land. All these factors mean that organic farming alone cannot produce enough food for us all. However, at present the UK is producing too much of the food we are able to grow here and farmers are being encouraged to 'set aside' fields where no crops are to be grown.

There are many arguments for and against organic farming, and for and against using synthetic chemicals in agriculture. There are no simple answers. Perhaps the most sensible solution is to encourage farmers to practise the natural, organic methods as far as possible.

Meanwhile, research chemists and biologists will carry on trying to develop fertilizers and pesticide chemicals which are effective, cheap and environmentally safe.

1 What sort of tests would you do to find out if there are any differences between 'organically-grown' food and food grown with the help of synthetic fertilizers?
2 If someone said to you: 'We've got to stop putting all these chemicals over our food crops – they are too dangerous', what would you reply?

3 The problems of feeding a hungry world

Everyone needs food. It is a basic requirement for human life. But you will know from television and newspaper reports that millions of people in the world do not have enough to eat and many die each year from starvation. Why does this happen? What are the problems involved in feeding the people of the world?

The rising population
Every year the world population increases by about 100 million (figure 17). That's an increase of more than the total population of the UK each and every year!

The rising population is likely to contribute to more famines and food shortages in the future. So efforts should be made to control its growth.

Some countries encourage families to have fewer children and have tried to increase the use of contraceptives. Better education is needed to persuade parents to have smaller families; and better health facilities also ensure that the small families survive.

Poverty
Many farmers in developing countries are not able to afford to buy machinery, seed of good quality, fertilizers or pesticides. They many also not be able to afford the technology necessary to irrigate their fields. They sometimes find themselves in a vicious circle where they have to sell foods to buy technology.

The poorest countries of the world are often heavily in debt. Instead of growing food they grow crops like tobacco or coffee to sell. The money they receive they use to pay

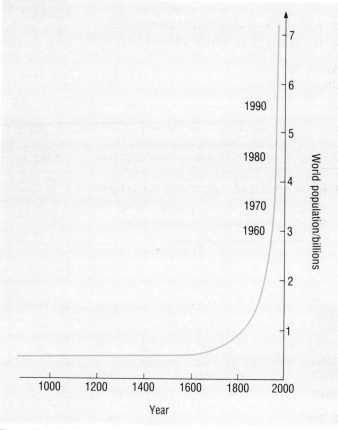

Figure 17 *World population changes*

their debts. The land which could be used for growing food is used to grow these 'cash crops'.

Harming the environment
In some parts of the world, such as Europe and North America, farmers make full use of their land by planting high-yielding seeds and applying fertilizers and pesticides. They also use machinery such as tractors to do much of the work.

This sort of intensive farming produces its own problems:

◆ soils may be over-used; they lose natural fertility and also become powdery as farmers use more and more synthetic fertilizers. In the grain fields of North America, this powdery soil is blown away and desert is being formed.

◆ intensive farming uses a lot of energy. Large amounts of non-renewable energy sources, such as petrol, gas and oil, are used in making fertilizers and pesticides and for farm machinery.

◆ rainforests are being destroyed to create grazing land for cattle. This is quickly over-used and the climatic changes caused by destroying the forests affects farmers everywhere.

At present there is enough food grown in the world to feed everyone. Rich countries produce too much food and poor countries do not produce enough. Distributing food from rich to poor will help solve immediate crises but are not long-term solutions.

Poor countries must have the capacity to feed their own people, control their own populations, use appropriate technology and develop education and health programmes.

1 Look at figure 19 showing how world population has increased. Why do you think the population has increased so much between 1880 and 1990? Why did the population size increase so slowly before this?

2 If the population increases at the same rate beyond the year 2000 what problems will the world face apart from food shortages?

3 What effect will future shortages of gas and oil have on farming in the UK?

In brief

Growing food

1 Plants consist of compounds made up of six main elements and smaller amounts of many others (figure 18).

3 To produce higher quantities of food farmers add extra fertilizer particularly to supply nitrogen (N), phosphorus (P) and potassium (K) (figure 20). Fertilizers can be classified as organic or inorganic; natural or synthetic.

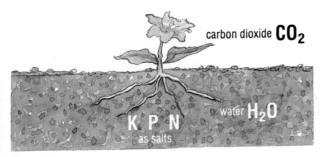

carbon dioxide CO_2

K P N as salts

water H_2O

Figure 18

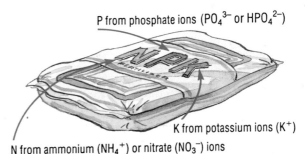

P from phosphate ions (PO_4^{3-} or HPO_4^{2-})

K from potassium ions (K^+)

N from ammonium (NH_4^+) or nitrate (NO_3^-) ions

Figure 20

2 Nitrogen taken out of the soil by plants can be replaced by three natural methods (figure 19).

plants take up nitrogen as nitrate ion NO_3^-

rain contains nitric acid

1 lightning causes nitrogen in air to react with oxygen

2 compost and manure can be added

3 bacteria in the soil convert N_2 gas into N compounds

Figure 19

4 Nitrogen is 'fixed' by reaction with hydrogen to form ammonia on a huge industrial scale in the Haber process. The temperature, pressure and catalyst for the process are chosen to make it as efficient as possible.

Source	Air (nitrogen and oxygen) Fractional distillation	Methane + Steam high temperature and pressure and catalysts
Reactants and product	nitrogen +	hydrogen $\rightleftharpoons$ ammonia 400°C 250 atm iron as catalyst
Equation	$N_2(g)$ +	$3H_2(g)$ $\rightleftharpoons$ $2NH_3(g)$

5 Ammonia is used to make nitric acid:

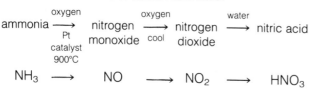

$$NH_3 \longrightarrow NO \longrightarrow NO_2 \longrightarrow HNO_3$$

6

> Ammonia
> ◆ is a gas at room temperature,
> ◆ is colourless,
> ◆ is very soluble in water,
> ◆ has a distinctive smell,
> ◆ forms an alkaline solution in water,
> ◆ turns litmus blue

7 Ammonia reacts with acids to form ammonium salts.

ammonia	+	nitric acid	→	ammonium nitrate
NH_3	+	HNO_3	→	NH_4NO_3

ammonia	+	sulphuric acid	→	ammonium sulphate
$2NH_3$	+	H_2SO_4	→	$(NH_4)_2SO_4$

8 All nitrates and ammonium salts are very soluble in water. Nitrate ions and, to a lesser extent, phosphate and potassium ions, are lost by 'leaching' out of the soil into streams and rivers.

Increasing levels of nitrate ions in drinking water are causing great concern as a health risk.

Excess nitrates and phosphates fertilize water plants and algae, which leads to eutrophication of lakes and slow-moving rivers.

9 Plant growth is affected by the acidity or alkalinity of the soil. This is indicated by pH measurement on the pH scale of 0 – 14, where 7 is neutral.

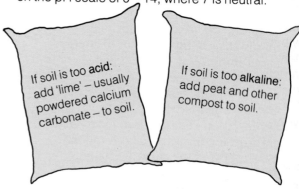

If soil is too **acid**: add 'lime' – usually powdered calcium carbonate – to soil.

If soil is too **alkaline**: add peat and other compost to soil.

10

Pesticides
— *insecticides* to kill *insects*
— *herbicides* to kill *weeds*
— *fungicides* to kill *fungi*

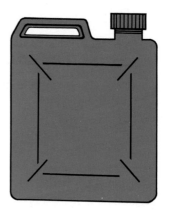

The ideal pesticide:

a) is very **selective** – kills or controls only the type of pest chosen,

b) has **no harmful effects** on humans or other animals or plants,

c) continues to be **effective for a long time** or does its work quickly then breaks down to harmless substances,

d) is **easy to store**, distribute and apply,

e) is **cheap**.

11 It is useful to compare the masses of atoms by their **relative atomic masses**. Adding up the relative atomic masses of atoms in a molecule gives the relative molecular mass of the molecule. Using these masses and knowing the balanced equation for a chemical reaction, it is possible to calculate the mass of products expected from known masses of reactants.

1 What chemicals are needed for life?

All living things must take in chemicals for survival. They need them for energy and for growth.

You obtain your chemicals by eating a variety of plants (fruits, vegetables, cereals) or animal products (milk, eggs, meat, etc.). You obtain oxygen by breathing and water by drinking.

Plants consist mainly of compounds such as carbohydrates (e.g. sugars and starch), proteins and smaller amounts of many other chemicals. Carbohydrates are made from the elements carbon, hydrogen and oxygen.

Table 1 *Nutrient elements essential for plants*

Element	Amount	Element	Amount
Carbon (C)	⎫	Iron (Fe)	⎫
Hydrogen (H)		Manganese (Mn)	
Oxygen (O)		Boron (B)	
Nitrogen (N)	LARGE	Copper (Cu)	TINY
Phosphorus (P)		Zinc (Zn)	
Potassium (K)		Molybdenum (Mb)	
Calcium (Ca)	⎭	Chlorine (Cl)	
Sulphur (S)	⎫ MEDIUM	Cobalt (Co)	⎭
Magnesium (Mg)	⎭		

Plants manufacture starch by the process of **photosynthesis**. All other essential elements are taken into plants from the soil through their root systems. These essential elements are known as **plant nutrients**. Elements which are only needed in tiny amounts are called **trace elements** or **micronutrients**.

Table 2 *Nutrients removed by crops*

Plant	Amount of nutrients removed by crops/kg per tonne of fresh material					
	N	P	K	Ca	Mg	S
Cereal (grain)	17.0	3.4	4.7	0.5	1.3	1.3
Cereal (straw)	6.0	0.7	6.8	3.0	0.8	0.9
Grass (hay)	14.0	2.6	15.0	3.4	1.0	1.0
Potato	3.0	0.4	4.8	0.2	0.2	0.3

From table 2, you will see that nitrogen and potassium are the elements most used up as grain, grass and potatoes are grown. Nitrogen is also lost:

a) by leaching of soluble nitrogen-containing compounds through the soil,

b) when the soil lacks oxygen, bacteria obtain oxygen by breaking down nitrate ions (NO_3^-) and release nitrogen as molecules of the gas (N_2).

If nitrogen and potassium and other elements are not in the soil, the plants will not grow properly. They must be added back to the soil. As farmers say – 'You must put back what you have taken out'.

Anything that is added to the soil to help plants to grow is called a **fertilizer**.

2 How is nitrogen put back into the soil?

Plants take nitrogen out of the soil. On farms, where large numbers of plants are being grown year after year, this nitrogen must be replaced. Farmers often do this by adding fertilizers. Yet there is an enormous natural source of nitrogen gas all around us – air. Nitrogen gas is four-fifths of the earth's atmosphere. Why do farmers buy costly fertilizers? Why don't they pump nitrogen gas directly into the soil to replace what is lost?

The answer comes from investigations of the chemistry of nitrogen. This gas exists in the form of N_2 molecules and these are remarkably resistant to reaction with other chemicals (figure 21). This unreactive gas has to be converted into a form which plants can use. This process is carried out in the manufacture of fertilizers. In fact the same effect can be achieved by nature itself.

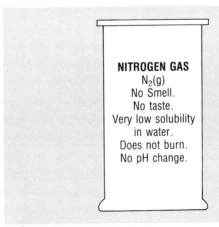

NITROGEN GAS
$N_2(g)$
No Smell.
No taste.
Very low solubility
in water.
Does not burn.
No pH change.

Figure 21

There are three natural ways in which nitrogen gas is converted into more usable, water-soluble compounds. These are called **nitrogen-fixing** processes. They provide natural fertilizers for the soil. Fields which farmers leave fallow after some years of use, will slowly regain their fertility from these natural sources.

a) Nitrogen-fixing bacteria

Large numbers of different bacteria live in the soil. Some of these, such as **azobacter**, are able to help convert nitrogen gas into proteins. The proteins can be broken down by other bacteria (decomposers) into compounds which are absorbed by plants through their roots.

Plants such as beans, peas, mustard and clover (all in the family called legumes) have a helpful relationship with bacteria called **rhizobium**. These bacteria live in small swellings (nodules) on the plant roots and consume some of the plant's store of carbohydrate (figure 22). In return, they convert nitrogen gas directly into useable forms for the plant. So effective is this process that farmers often sow a field with mustard, then they plough it in, to provide nutrient nitrogen for other plants sown later. This type of mutual help between bacteria and root nodules is an example of **symbiosis**.

b) Nitrogen in a lightning flash

Lightning releases huge amounts of energy (figure 23). This helps nitrogen and oxygen in the air to react together to form nitrogen monoxide (NO). Nitrogen monoxide then reacts with oxygen in the air to make nitrogen dioxide (NO_2), which in turn reacts with water in the atmosphere to make nitric acid (HNO_3). Nitric acid supplies nitrate ions (NO_3^-), to the soil and these are absorbed by the plant roots.

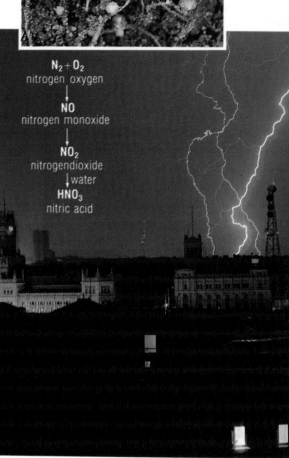

◀ **Figure 22**
Nitrogen-fixing nodules on the roots of French beans

$N_2 + O_2$
nitrogen oxygen
↓
NO
nitrogen monoxide
↓
NO_2
nitrogendioxide
↓ water
HNO_3
nitric acid

Figure 23   Lightning releases large amounts of energy into the atmosphere

c) Organic fertilizers
('You can't beat a good dollop of muck')

For centuries, farmers and gardeners have returned nutrients to the soil using **compost** and **manure**. Compost is well-decayed plant material and manure is mainly animal excreta plus straw (figure 24). These are often described as organic fertilizers: the term 'organic' here means that the substances have come from living or once-living plants or animals.

Both compost and manure are excellent sources of nitrogen and other nutrients, including trace elements (figure 25). They also act as soil conditioners by admitting air and water and improving drainage in heavy soil and giving 'body' to light sandy soils.

Figure 24 ▲
Exothermic reactions inside this dung heap are causing it to steam.

Figure 25   Well-rotted garden compost is an invaluable addition to any soil.

3 How was the nitrogen problem fixed?

Natural methods of nitrogen fixation are often not quick enough to supply the fields with sufficient nitrogen compounds for modern methods of intensive farming. Every day, over 60 million tonnes of nitrogenous fertilizers are made. This would have seemed like a miracle to farmers in the early part of this century. At that time the natural sources, whether organic or inorganic, were already not keeping up with demand. Yet by 1913, the 'nitrogen problem' was solved.

Figure 26 *Fritz Haber*

Figure 27 *Carl Bosch*

Brilliant theoretical work by the chemist Fritz Haber (figure 26) and the practical genius of engineer Carl Bosch (figure 27) found a way for the comparatively cheap 'fixation' of nitrogen from the atmosphere. Their solution was the **Haber Process** – the production of ammonia gas on an industrial scale, by direct reaction between nitrogen and hydrogen.

$$\text{nitrogen} + \text{hydrogen} \rightleftharpoons \text{ammonia}$$
$$N_2(g) + 3H_2(g) \rightleftharpoons 2NH_3(g)$$

This reaction is **reversible**, as ammonia, when heated, decomposes back into nitrogen and hydrogen. After a while, ammonia decomposes as quickly as it is formed. When the rates of the opposite reactions are balanced, the reaction is said to be at **equilibrium**.

The main problem facing Haber was: how could he ensure that ammonia was formed faster than it decomposed, i.e. how could he prevent the reaction from reaching the equilibrium? He found that removing the ammonia as it is formed and careful adjustment of the temperature and pressure gave the best conditions. A catalyst speeded up the whole process. Bosch solved the huge engineering problems that Haber's solution created.

A typical modern Haber Process plant produces 1200 tonnes of ammonia each day. It runs at about 400°C and a pressure of 200 atmospheres. The catalyst is powdered iron packed in small hard pellets. The way in which the iron acts as the catalyst for this reaction is described on page 186.

4 How is ammonia converted into a fertilizer?

The Haber Process is an important stage in the fixation of nitrogen, as the main use of ammonia is in the fertilizer industry. The ammonia is first converted into nitric acid, then from nitric acid, nitrates can be made, such as **ammonium nitrate** (a nitrogenous fertilizer).

nitrogen	→	ammonia	→	nitrogen monoxide	→	nitrogen dioxide	→	nitric acid
N_2	→	NH_3	→	NO	→	NO_2	→	HNO_3

There are some similarities and some differences between this route from nitrogen to nitric acid and those described on page 96. The essential differences are in the conversion of nitrogen to nitrogen monoxide (figure 28). Huge quantities of energy are required for the direct conversion and this was supplied by lightning. Unless very cheap electrical power is available, it is cheaper to convert nitrogen to ammonia and ammonia to nitrogen monoxide using a catalyst (figures 29 and 30 on the next page).

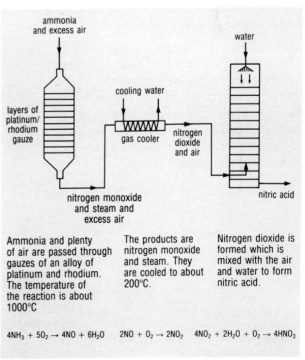

Ammonia and plenty of air are passed through gauzes of an alloy of platinum and rhodium. The temperature of the reaction is about 1000°C

The products are nitrogen monoxide and steam. They are cooled to about 200°C.

Nitrogen dioxide is formed which is mixed with the air and water to form nitric acid.

$4NH_3 + 5O_2 \rightarrow 4NO + 6H_2O$ $2NO + O_2 \rightarrow 2NO_2$ $4NO_2 + 2H_2O + O_2 \rightarrow 4HNO_3$

Figure 28

Figure 29 *The rock – like catalyst used in the Haber process.* ▶

Figure 30 *A Haber Process plant producing ammonia for ICI at Billingham.* ▼

Much of the nitric acid is converted into solid fertilizers such as ammonium nitrate.

$$NH_3 \quad + \quad HNO_3 \quad \rightarrow \quad NH_4NO_3$$

5 What is in a bag of fertilizer?

For many years farmers have used synthetic fertilizers, such as the one shown in figure 31. It contains a mixture of compounds of nitrogen, phosphorus and potassium.

Figure 31 *This bag of NPK fertilizer adds nutrients to the soil in the ratio of 20 parts nitrogen, 8 parts phosphorus, and 14 parts potassium.*

Nitrogen (N) is converted into several compounds called **nitrogenous fertilizers**. These include compounds which contain the ammonium ion (NH_4^+ as in ammonium nitrate (NH_4NO_3)). Another important nitrogenous fertilizer is urea ($CO(NH_2)_2$).

The nitrate ion is the most direct source of nitrogen for plants as their roots absorb it easily. Ammonium ions and compounds such as urea are oxidised by soil bacteria to nitrate ions, before they are absorbed by the roots of plants.

Phosphorus (P) is usually present as compounds containing phosphate ions (PO_4^{3-} or HPO_4^{2-}). The manufacturer does not give the percentages of phosphorus present. Instead we are told the percentage of phosphorus oxide (P_2O_5). This is rather confusing as P_2O_5 is not actually present in the fertilizer. There is no scientific reason for mentioning P_2O_5! Britain's farmers depend upon imported calcium phosphate rock ($Ca_3(PO_4)_2$) for their main supplies of phosphate fertilizers.

Potassium (K)   The most important potassium compound used is potassium chloride (KCl). This gives potassium ions (K^+). In Britain the largest source is in North Yorkshire, where Sylvinite, a mixture of potassium chloride and sodium chloride, is mined from layers 10 metres thick. The potassium and sodium chlorides can be separated by fractional crystallisation.

Again, you will see on the bag in figure 31 that the amount of potassium is quoted as the percentage of its oxide (K_2O). Like P_2O_5, this is not actually present in the fertilizer.

6 Why do fertilizers cause problems?

Fertilizers help people to grow more food. But there is, unfortunately, another story to tell (figure 32).

Nitrate pollution threat!

Algal blooms in Venice caused by leaching of fertilizers

Farmers produce record crops EC food mountains higher than ever

Figure 32

Many of the fertilizers used today contain nitrate ions or compounds which are converted to nitrate ions in the soil. This works well for plants but, if nitrate ions get into the water supply, they may be harmful. They can be converted in your digestive system to nitrite ions. These ions interfere with the way your blood carries oxygen around your body.

The presence of nitrates in water causes another problem. Nitrates and phosphates (as fertilizers or detergents) when washed into rivers, lakes and streams fertilize plants living in them. This encourages the growth of the water-plants which then choke the water course (figure 33). The most damaging are the tiny plants called **algae**. These grow so thickly in great, green, cloudy masses called 'algal blooms' that sunlight is cut off from the other water-plants. The plants die and the algae too are short-lived. Bacteria and fungi cause these plants to decompose and so use up the dissolved oxygen in the water. Eventually all the oxygen is used up and then all the life in the water – fish, snails, larvae, the bacteria themselves – dies. This process is called **eutrophication** of the water. In the worst cases it leaves the river or lake lifeless and stinking.

Figure 33   *This pond is choked by algae and dead twigs.*

It is not enough simply to blame synthetic fertilizers for the large amounts of nitrate in the water. This can come from many sources – from manure, sewage and compost; from plants such as clover or peas; or from the oxidation of nitrogen compounds following the ploughing up of grasslands.

Much more care all round from farmers and Water Authorities, with improved sewage disposal, is needed for the health of all water-courses.

7   *Importance of knowing the pH of the soil*

Farmers and gardeners are always concerned about the acidity or alkalinity of their soil. The scientific measure of acidity and alkalinity is the pH scale of 0 – 14, with 7 as the neutral point (figure 34). Pure water has a pH of 7.

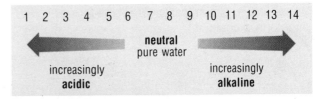

Figure 34   *The pH scale*

Many plants prefer a soil pH which is slightly acidic, around 6.5. Some will tolerate higher acidity (lower pH value), others thrive in more alkaline soils (higher pH values) (figure 35).

Figure 35 ▲
This Somerset peat is acidic and is cut and sold to growers in areas with alkaline soils. Why?

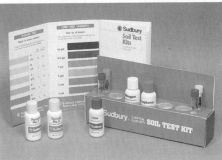

Figure 36   *A soil pH testing kit. How do you think it works?*

Soil with a high peat content or with minerals such as iron compounds or rotting vegetation and lack of oxygen, tend to be acidic, perhaps as low as pH 4.

Soils in limestone or chalky areas are alkaline – up to pH 8.3. Farmers and gardeners measure the pH of the soil to see whether or not it will suit the needs of particular plants.

Changing the pH of soil: liming

The most common problem for farmers in many parts of Britain is soil that is too acidic. Soil bacteria do not like this condition. Substances are needed which will react with the acids in the soil in a **neutralisation reaction**. They must be available in very large quantities, easy to transport, safe to use and cheap. Fortunately Britain has vast supplies of the ideal compound – calcium carbonate in the form of **limestone** or **chalk**.

◀ **Figure 38** Liming fields is important in areas with acid soils. Why?

Figure 37 *Modern lime kilns*

In past centuries limestone or chalk were usually roasted strongly in lime kilns (figure 37) to convert them into the more easily powdered and reactive alkali, calcium oxide (commonly called 'quicklime').

$$\text{calcium carbonate} \xrightarrow{\text{heat}} \text{calcium oxide} + \text{carbon dioxide}$$

$$CaCO_3 \xrightarrow{\text{heat}} CaO + CO_2$$

Calcium oxide reacts vigorously with water to give calcium hydroxide ('slaked lime').

$$\text{calcium oxide} + \text{water} \rightarrow \text{calcium hydroxide}$$

$$CaO + H_2O \rightarrow Ca(OH)_2$$

All these substances, calcium carbonate, calcium oxide and calcium hydroxide, may be used to reduce soil acidity.

$$\text{acid} + \begin{cases} \text{calcium carbonate} \\ \text{or} \\ \text{calcium oxide} \\ \text{or} \\ \text{calcium hydroxide} \end{cases} \rightarrow \text{calcium salt} + \text{water}$$

Farmers call them all 'lime' when they are 'liming' their fields (figure 38). In modern farming the 'lime' used is almost always powdered chalk or limestone (calcium carbonate). It has low solubility in water and does not wash away as quickly as the other

Figure 39 *The quarrying of chalk or limestone creates huge pits, like this one in Kent. Approximately 4 million tonnes are needed to improve acid soils each year and it has many other uses.*

compounds. As well as reducing soil acidity the compounds of 'lime' supply calcium ions (Ca^{2+}) to the plants. They also help to improve the general condition of the soil.

Soil in chalk and limestone areas of Britain tends to be alkaline. For some crops to grow well, the pH must be lowered. It helps to dig in plenty of peat and decaying organic material, such as compost or manure, as these can be fairly acidic.

8   How much do we need?

An order for 8000 tonnes of ammonium nitrate arrives at the fertilizer factory. It is important that there is no wastage. So how much ammonia and nitric acid are needed?

The symbol equation shows the number of particles of each reactant and product.

$$NH_3 + HNO_3 \rightarrow NH_4NO_3$$

This shows you that **one** molecule of ammonia reacts with **one** molecule of nitric acid to make **one** molecule of ammonium nitrate.

To use this information, you also need to know the **relative masses** of atoms of hydrogen, nitrogen and oxygen.

It is known that:

1 atom of nitrogen weighs **14 times** more than 1 atom of hydrogen and 1 atom of oxygen weighs **16 times** more than 1 atom of hydrogen.

But how will this help the manager of the fertilizer factory to work out how much ammonia and nitric acid are needed?

One molecule of ammonia (NH_3), contains one nitrogen atom and three hydrogen atoms. Thus the **relative molecular mass** of ammonia is:

$$\boxed{N} \quad \boxed{H_3}$$
$$14 \quad + \quad (1 \times 3)$$
$$= 17$$

The relative molecular mass of nitric acid (HNO_3) is:

$$\boxed{H} \quad \boxed{N} \quad \boxed{O_3}$$
$$1 \quad + \quad 14 \quad + \quad (16 \times 3)$$
$$= 63$$

The relative molecular mass of ammonium nitrate is:

$$\boxed{N} \quad \boxed{H_4} \quad \boxed{N} \quad \boxed{O_3}$$
$$14 + (1 \times 4) + 14 + (16 \times 3)$$
$$= 80$$

Then:

	NH_3	+	HNO_3	=	NH_4NO_3
Relative Molecular Mass	17		63		80

To make 80 grams of ammonium nitrate, you need 17 grams of ammonia and 63 grams of nitric acid.

So to make 8000 tonnes of ammonium nitrate, you need 1700 tonnes of ammonia and 6300 tonnes of nitric acid.

Chemists always use the relative atomic masses and relative molecular masses when they wish to calculate how much reactant they need and how much product they hope to form. It is important that you understand this idea. To help you, here are some more examples.

Example 1 100 tonnes of limestone (calcium carbonate) are dug from a quarry. If all of it is heated in a lime-kiln:
a) how much lime (calcium oxide) is produced?
b) how much carbon dioxide is produced?

The balanced equation for the reaction is:

$$CaCO_3 \quad \rightarrow \quad CaO \quad + \quad CO_2$$

We need to know that the relative atomic masses of calcium, carbon and oxygen are 40, 12 and 16.

The answers to a) and b) are 56 and 44 tonnes.

Example 2 In photosynthesis, carbon dioxide and water are converted into glucose.

$$6CO_2 \ + \ 6H_2O \ \rightarrow \ C_6H_{12}O_6 \ + \ 6O_2$$

How many grams of glucose are produced from 72 grams of carbon dioxide?

We know that the relative atomic masses of carbon, oxygen and hydrogen are 12, 16 and 1.

The answer to the question is 49 grams.

Example 3 The formula of the herbicide 2,4,5-T is $C_8H_5O_3Cl_3$. What is the percentage of carbon in 2,4,5-T?

The relative atomic masses of carbon, hydrogen, oxygen and chlorine are 12, 1, 16 and 35.5.

The answer to the problem is 37.6%.

Taking it further

In the calculations above, you have only needed the *relative* masses of atoms. For example, you needed to know that one atom of oxygen has a mass that is 16 times that of one atom of hydrogen. This means that we need one element as the reference point and then use the mass of the atoms of all the elements relative to that one. Carbon has been chosen and given the relative atomic mass of 12. One atom of carbon has 12 times the mass of one atom of hydrogen.

The relative atomic masses of some common elements are:

hydrogen	carbon	nitrogen	oxygen	sodium
H	C	N	O	Na
1	12	14	16	23

aluminium	silicon	phosphorus	sulphur	chlorine
Al	Si	P	S	Cl
27	28	31	32	35.5

potassium	calcium	iron	copper
K	Ca	Fe	Cu
39	40	56	64

These are *relative* masses. The actual mass of a hydrogen atom is very small – about 2×10^{-24} grams.

0.00000000000000000000000002g

Things to try out

1 *Make fertilizer* Make your own ammonium phosphate fertilizer from rock phosphate. (See 'Chemicals from Agriculture', No 5 in Experimenting with Industry, published by ASE).

2 *Comparing lime* Design and try out experiments to compare how well different types of 'lime' (calcium carbonate (limestone), calcium oxide (quicklime), calcium hydroxide (slaked lime)), reduce acidity in soil. If naturally acidic soil is not available you could make some by adding a little vinegar to your local soil.

Things to find out

3 a) Find out why DDT and compounds like it are called 'organochlorines' by chemists.
   b) Find out more about methods of reducing the number of insects without using insecticide chemicals, and find examples of 'pheromones' and how they are used.
   c) Find out how bacteria and viruses are used to 'fight bugs with bugs'.

4 Find out where Britain's supplies of calcium phosphate rock come from. What would happen if these supplies became unavailable?

5 a) Find out how nitrate ions are detected and measured in water supplies.
   b) Why is it very difficult and expensive to reduce nitrate levels in streams and rivers?

6 Find out more about agriculture as a source of chemicals other than food, e.g. fuels, medicines, etc.

7 Read the information and instructions on a packet or bottle of insecticide or weedkiller. Do you understand them fully? Do other members of your family and friends understand them? If not write a letter which tells the manufacturer that you are not satisfied with the information and instructions and offer suggestions for improvement.

8 Find out from a library or local government office:
   a) approximately how many weeks supply of food is supposed to be kept stored in Britain for our own population in case of emergency,
   b) how large quantities of food such as grain, meat and milk products are stored,
   c) what the difficulties and costs are of storing these foods.

Points to discuss

9 Who should pay the main costs of reducing pollution (e.g. by nitrates and phosphates) in our rivers? Some say it should be the farmers; others say that everyone needs the food the farmers grow and so all should pay. What do you think?

10 In which countries is there a famine at present? Choose one country and discuss in what ways science and technology could help to relieve that famine. What assistance could Britain give?

Questions to try

11 In the Haber Process:
   a) why is the pressure used about 200 atmospheres rather than 1000 atmospheres or 10 atmospheres,
   b) why is the temperature about 450°C rather than 1000°C or 100°C,
   c) why is the main catalyst powdered iron, rather than platinum which is more effective,
   d) how is ammonia removed from the process and how is it stored?

12 Complete the following word equations:

ammonium chloride $\xrightarrow{\text{heat}}$
ammonium nitrate $\xrightarrow{\text{heat}}$
ammonia +   → ammonium phosphate +
calcium carbonate + hydrochloric acid →
ammonium sulphate + calcium hydroxide →

13 Below are some compounds containing nitrogen:

ammonia (NH_3)   urea ($CO(NH_2)_2$)
sodium nitrate ($NaNO_3$)   nitrogen dioxide (NO_2)

(Relative atomic masses are H 1, C 12, N 14, O 16, Na 23.)

   a) Calculate the percentage by mass of nitrogen in each compound.
   b) Which of the compounds should be the most effective nitrogenous fertilizer?
   c) Which one is actually used most widely in Britain and why?
   d) Which one is never used as a fertilizer? Explain why.

14 If you were given three gas jars containing different gases and told they could be carbon dioxide, nitrogen, oxygen or ammonia, how would you decide which gas is in each jar?

Introducing food processing

The very earliest humans ate fresh food all the time but they soon discovered that fresh food quickly goes bad. They found ways of **preserving** food by cooking it, drying it and salting it. They discovered that adding spices to food improved the flavour of the food as well as preserving it. They learnt that, by mixing different foods together, they could get better tastes and textures.

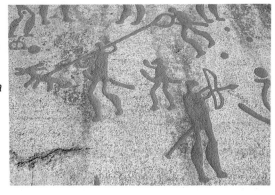

Figure 1 *The first humans were hunters. If they caught a large animal, they could not eat it all at once. How did they preserve it?* ▶

Figure 2

Figure 3

Which of the foods in **figures 2-5** *are completely unprocessed?* ▶

Figure 4

Figure 5

These were the beginnings of **food processing** – methods of treating food to improve it and make it last longer. Nowadays, most of the food you eat is processed in one way or another. Some people say too much processed food is eaten in this country.

Chemistry has made it possible to preserve, colour, flavour and improve food in many ways. As a result, there is an enormous variety of different foods to choose from all the year round.

In this chapter you will see how

◆ chemistry helps to preserve, process and improve food,
◆ chemistry is used to find out what is in food,
◆ an understanding of chemistry helps you to understand the properties of food.

▲ **Figure 6** *Crisps are made from potatoes. What processing is done to preserve the crisps and make them tasty?*

Figure 7 *Many different flavours have been used to make the wide variety of sweets in this shop. When you eat a sweet, how can you tell which fruit flavour it is supposed to be?* ▶

1 Preserving food

The problem with food is that it 'goes off' – goes bad. If you live on a farm, it may be possible to get fresh food all the time. However, most people need to *preserve* food is some way, to stop it going off before they can eat it.

People have been preserving food for thousands of years. Drying, pickling and salting are the traditional ways of preserving food and these methods are still used today.

Why does food need preserving?

Humans are not the only living things which need food. **Microbes** – bacteria and fungi – eat our food too. Unless they are stopped, they grow in food and make it go bad. Some microbes produce harmful substances which cause food poisoning. All food processing methods aim to stop the growth of microbes.

One way of preventing microbes growing is to kill them. This process is called **sterilisation**. It is used when food is **canned** or **irradiated** (see table 1).

Another method is to stop the microbes reproducing so quickly. This can be done by **drying** or **freezing** the food, so the microbes do not have enough water.

Salt or acids (like pickling vinegar) can be added to the food. These stop microbes growing.

Table 1 summarises some of the important methods used to preserve food.

▲ **Figure 8** *Dried foods like these will keep for many years. Drying is one of the commonest ways of preserving food.*

▲ **Figure 9** *Pickles*

◀ **Figure 10** *Uncooked meat quickly goes off. Burgers contain chemical preservatives, like salt and sodium nitrite, to make them keep longer.*

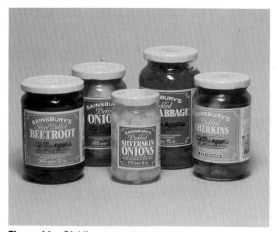

Figure 11 *Pickling vegetables in vinegar preserves them for many months.*

Look carefully at table 1 and then answer these questions.

1 Which of the methods in the table is used to preserve each of the following:
ice cream, yeast extract (e.g. *Marmite*), raisins, spaghetti, marmalade, chutney?

2 Suppose you wanted to preserve some apples. Which of the methods in the table might you use?

3 Which of the methods in the table keeps food closest to its original taste and texture?

4 Imagine setting off on a long sea voyage, two hundred years ago. Which methods could be used to preserve the food for your journey?

Table 1 *Some of the important methods used to preserve food*

Method	What is done	How it works	Examples
Canning	Food is cooked and then sealed in cans	Cooking kills microbes. Sealing in cans stops any more getting in	All kinds of tinned foods, like soups, meat, beans, etc.
Drying	Food is dried in the sun or in special ovens or freeze-driers	Drying takes away the water needed by microbes	Dried beans, lentils, peas. Dried fruit
Freezing	Food is quickly frozen	Freezes the water needed by microbes. Slows down biochemical changes	Frozen fish fingers, burgers, peas, etc.
Pickling in oil	Food is cooked and then covered in spiced oil	Cooking kills microbes. Oil stops any more getting in. Spices prevent microbes growing	Mango pickle, lime pickle
Pickling in vinegar	Food is covered in vinegar and bottled	Microbes do not grow well in acid conditions	Pickled onions, pickled beetroot
Chemical preservatives	Chemicals which control microbes are added to food	The chemicals are poisonous to microbes but safe for humans	Beefburgers contain sodium metabisulphite (E223) as a preservative
Salting	Food is soaked in a strong salt solution	Microbes do not grow well in salty conditions	Salted fish, bacon
Sugaring	Food is kept in a very strong sugar solution	Microbes do not grow well in highly sugary solutions	Jam
Irradiation	See question 7 in Things to do		

2 What are food additives?

For centuries, substances have been *added* to food to preserve it or to improve it. Thousands of years ago the Egyptians used food colourings and spices for flavouring. Today the use of food **additives** is very common and some people are worried that these additives may be a danger to health.

The labels on most foods now show which additives the foods contain. Figure 12 shows an example.

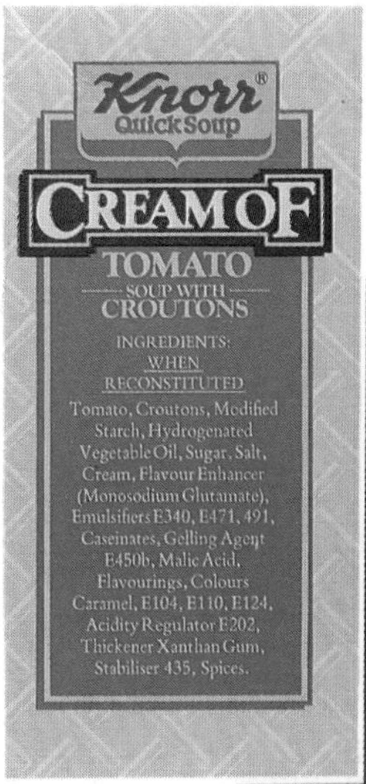

◀ **Figure 12** *List the food ingredients and additives given in this ingredients list for a packet of dried soup mix.*

Figure 14 ▶
This diet drink contains artificial sweeteners instead of sugar, so it is less fattening.

Figure 13 ▶
When peas are processed (as in the tin on the right) their colour alters and artificial colouring is added to make them look like fresh peas.

Most food additives are **government-regulated**. This means the government has given permission for the additive to be used. These regulated additives each have a number. If the number begins with an **E**, it means the additive is also regulated by the **EC** (European Community).

Table 2 *Important types of food additives*

Type of additive	Additive numbers used	Job done by additive in food	Example
Colourings	most begin with 1	give food attractive colour or replace colour lost in processing	green colouring (E142) added to tinned peas
Preservatives	most begin with 2	preserve food so it goes off less quickly (see page 104)	sulphur dioxide (E219) used to preserve fruit juice
Flavourings	not numbered	give flavour to food	fruit flavours used in ice lollies
Anti-oxidants	300 to 321	stop fats getting oxidised and tasting nasty	butylated hydroxytoluene (E321) used in potato crisps
Emulsifiers and stabilisers	E322 and some numbers between E400 and E495	make oil and water mix (see page 114)	lecithin (E322) used in powdered milk
Acids and bases	most begin with 5	various, including control of pH	sodium hydrogencarbonate (500) in tinned custard
Sweeteners	most begin with 4 or 6	sweeten food without using sugar	sorbitol (E420) in diabetic chocolates

There are a number of different types of additives. Each type does a different job. Table 2 shows the most important types.

Look carefully at table 2 and then answer these questions.

1 Which types of additives might be put in each of the following:
orange squash drink, margarine, potato rings snack food, jam?

2 A cheese spread contains the following additives:
E160, E202, E320
What job does each additive do?

Are additives safe?

No food additive can be used unless it has been tested and approved. The additives are tested on animals to try to make sure they are not dangerous to human health. Additives that are regulated by the government have numbers. Flavourings are not regulated in the same way, although they do have to be tested and approved. Most flavourings are a mixture of many different substances and often they are a trade secret.

As a result of safety testing, some food additives that used to be permitted are now banned. For example, in 1957 there were 30 colouring additives permitted in Britain. Today only half of these additives are still allowed, although others have since been added to the list.

In spite of all the testing, many people are worried that some additives may be unsafe. Some of the possible dangers are

◆ *Allergies*   Certain people may be allergic to particular additives. They may make them come out in a rash, for example, or cause stomach upsets.
◆ *Hyperactivity*   Hyperactive children are over-active and do not sleep much. It is possible that hyperactivity is caused by certain additives, particularly colourings.
◆ *Long-term illness*   It is very difficult to find out whether additives cause long-term illnesses, such as cancer, because the tests take so long to carry out. Certain additives have been banned because they cause cancers when fed to animals for long periods.

Figure 15   *These highly-processed foods require the addition of flavour additives so that they taste as expected.* ▶

1 Would it be possible for all food to be completely free from additives?

2 Susan is aged 3. She sleeps only three hours a night and is over-active by day. Her doctor suspects that her hyperactivity is caused by the yellow food colour **Tartrazine** (E102). What should her parents do?

3 Suppose there was a government ban on *all* flavouring additives. How would this affect the kind of food *you* eat? Which kinds of foods would be affected most?

4 Some of the colourings used by food companies are natural colours. For example, **cochineal** is a red dye made from tropical insects. **Carotene** is an orange dye made from carrots. Do you think natural dyes are likely to be safer than dyes which are made artificially by chemists? Give your reasons.

3 *Chemistry ripens bananas*

British people have a liking for tropical fruits such as bananas, peaches and avocados. The problem is, these fruits have to travel a long way from the countries where they are grown. They have to be brought in unripe, otherwise they would be over-ripe by the time they were sold.

The fruit has to be ripened after it has arrived. How do you get fruit to suddenly ripen after weeks of deliberately keeping it in an unripe condition?

Two things are done to help this ripening process. The fruit is put in a warm place. It is exposed to small amounts of **ethene** gas, which triggers off the ripening process.

Ethene (C_2H_4) is the simplest alkene. It is produced naturally by fruit as it ripens and it acts as a trigger to start off the ripening of other fruit. Releasing small amounts of ethene into a store room of unripe fruit helps it ripen faster. Once the fruit has started ripening, it gives off its own ethene.

The unripe fruit is put into large, warm rooms. Ethene is made by a chemical reaction which involves passing alcohol over a heated catalyst. Only small amounts of ethene are made. Too much ethene would cause a fire risk and would in any case be bad for the fruit. If the amount of ethene in the ripening room gets too high, the excess is removed by reacting it with potassium permanganate.

Figure 16 .*The controlled emission of ethene in this banana warehouse ripens the fruit.*

1 What changes would you notice in potassium permanganate as it absorbed the excess ethene?

2 Tomatoes can be grown in Britain but in bad summers they often stay green. They fail to ripen to a red colour. A good way of ripening green tomatoes is to put them in a closed box with one or two *ripe* tomatoes. How does this work?

3 Do you think there would be any difference in taste between fruit which is ripened artificially using ethene and fruit which has ripened naturally on the plant?

1 **Raising agents** are used to produce bubbles of carbon dioxide to make cakes rise. Sodium hydrogencarbonate is often used as a raising agent. It is used on its own or mixed with tartaric acid.

2 Carbonates and hydrogencarbonates give off carbon dioxide and produce a salt and water when they are reacted with acid.

3 All hydrogencarbonates decompose on heating, forming carbonates, water and carbon dioxide. Carbonates decompose on heating to give a metal oxide and carbon dioxide. The less reactive a metal, the easier it is to decompose its carbonate. Carbonates of very reactive metals are stable when heated.

4 Figure 17 shows some of the properties of food that change when it is cooked.

Figure 17 *Changes take place in food when it is cooked*

5 Some of the vitamin C in a food is destroyed when the food is cooked.

6 You can find how much vitamin C is in a food by titrating it with DCPIP.

7 Figure 18 shows two factors which affect how quickly apples and other fruit go brown.

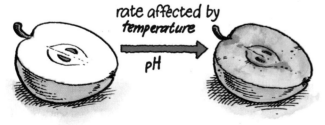

Figure 18

8 Oils and fats have similar chemical compositions. Oils are less saturated than fats. Oils can be turned to fats by hydrogenation. Hydrogenation is a reaction in which hydrogen is added to another substance.

9 Margarine is made by blending fats and oils with milk.

10 The amount of **unsaturation** in a fat or oil can be tested using bromine water or aqueous potassium permanganate.

11 **Miscible** liquids mix together completely and do not form layers. **Immiscible** liquids do not mix so they form layers.

12 Figure 19 summarises key ideas about emulsions and emulsifying agents.

Oil and water do not mix. They are immiscible.   Add emulsifying agent. Stir.   Oil and water mix, forming an emulsion.

Figure 19

13 Many foods are emulsions. Emulsions are either oil droplets suspended in water or water droplets suspended in oil. Milk is an emulsion of oil droplets suspended in water. Butter is an emulsion of water droplets suspended in oil. Milk can be turned to butter by shaking it with air.

14 Food additives are chemicals which are added to food to preserve or improve it. Food additives need careful testing to check they are safe. Figure 20 shows the main classes of additives.

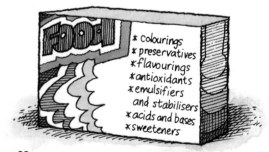

* colourings
* preservatives
* flavourings
* antioxidants
* emulsifiers and stabilisers
* acids and bases
* sweeteners

Figure 20

15 Food colourings are a mixture of dyes. They can be separated by **chromatography**.

16 Flavourings are used to improve the taste of food. They are also used to make up for flavour which is lost when the food is processed.

17 **Esters** are compounds made by combining an acid and an alcohol. Esters are present in many fruit flavourings.

How can you make cakes rise?

Food usually tastes nicer if it is light and airy, rather than heavy and solid. Take bread, for instance. Many people find bread much nicer to eat if it has 'risen' than if it is solid and **unleavened** (figure 21).

Figure 21 *The white loaf on the right is well risen and is full of bubbles. The unleavened bread on the left is much heavier and does not contain many bubbles.*

Cooks make food light and airy by mixing together ingredients which will produce tiny bubbles of gas. The gas produced is usually carbon dioxide. Figure 22 shows how this works with a cake.

Figure 22 *How a cake rises*

How is this carbon dioxide actually produced in the making of the cake? There are two chemical reactions which can be used to produce the carbon dioxide:

◆ reacting sodium bicarbonate with an acid,
◆ heating sodium bicarbonate.

How do bicarbonates and carbonates react with acid?
Bicarbonates are compounds containing the HCO_3^- ion, called the hydrogencarbonate ion. Chemists prefer to call bicarbonates **hydrogencarbonates**. From now on we shall call sodium bicarbonate ($NaHCO_3$) by its correct chemical name – **sodium hydrogencarbonate**.

Carbonates are compounds containing the carbonate ion (CO_3^{2-}). For example, sodium carbonate is Na_2CO_3.

If you add acid to a carbonate or hydrogencarbonate, it fizzes vigorously. Carbon dioxide is given off.
  All carbonates and hydrogencarbonates react with acid to give carbon dioxide, water and a salt.

For example:

sodium carbonate	+	hydrochloric acid	→	carbon dioxide	+	water	+	sodium chloride
Na_2CO_3	+	$2HCl$		CO_2	+	H_2O	+	$2NaCl$

These reactions are very important. For example, limestone is calcium carbonate and it reacts slowly with acids in the rain. This is why limestone buildings are attacked by acid rain (see page 155).

One way of making cakes rise is to use a mixture of sodium hydrogencarbonate and tartaric acid, called baking powder. Both these substances are safe to eat. They react together in the cake mixture to make bubbles of carbon dioxide.

Figure 23 *Sherbet contains a mixture of sodium hydrogencarbonate and citric acid. When these substances dissolve in the liquid in your mouth, they produce carbon dioxide - they FIZZ! (See page 117 for recipe.)*

What happens when you heat carbonates and hydrogencarbonates?
Figure 25 shows an experiment to investigate what happens when you heat carbonates and hydrogencarbonates. Table 3 shows some results.

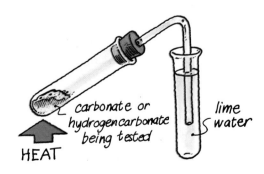

Figure 24 *Testing the effect of heat on carbonates and hydrogencarbonates.*

Table 3   *Results of heating carbonates and hydrogencarbonates.*

Substance being heated	Result
sodium hydrogencarbonate	Gas given off rapidly. Lime water turns milky.
potassium hydrogencarbonate	Gas given off rapidly. Lime water turns milky.
copper carbonate	Gas given off rapidly. Lime water turns milky.
lead carbonate	Gas given off rapidly. Lime water turns milky.
potassium carbonate	Gas given off slowly.
sodium carbonate	Gas given off slowly.
zinc carbonate	Gas given off steadily. Lime water turns milky.

1 What gas is given off when carbonates and hydrogencarbonates are heated? How do you know?

2 Which decomposes more easily – a carbonate or a hydrogencarbonate? How do you know?

3 Place the carbonates in order of how readily they give off the gas when heated.

The general rules about the effect of heat on carbonates and hydrogencarbonates are:

◆ All hydrogencarbonates are easily decomposed by heat, forming carbon dioxide, water and a carbonate. For example:

sodium hydrogencarbonate	→	carbon dioxide	+	water	+	sodium carbonate
$2NaHCO_3$	→	CO_2	+	H_2O	+	Na_2CO_3

◆ Most carbonates are decomposed by heat, forming carbon dioxide and an oxide. For example:

copper carbonate	→	copper oxide	+	carbon dioxide
$CuCO_3$	→	CuO	+	CO_2

The carbonates of more reactive metals like sodium and potassium are more difficult to decompose.

When calcium carbonate (limestone) is heated, it decomposes to form calcium oxide (lime). Farmers use both to neutralise acidity in the soil (page 100).

Cakes can be made to rise by adding sodium hydrogencarbonate to the cake mixture. When the cake is cooking, the sodium hydrogencarbonate decomposes. This produces bubbles of carbon dioxide which make the cake mixture rise.

2   Measuring vitamin C

Vitamin C is an important part of our diet. Lack of vitamin C can cause the disease **scurvy**. Some scientists believe vitamin C helps to protect us against many other diseases. You can find the amount of vitamin C in a food (as in figure 25) by titrating it with a pink dye called DCPIP. When it reacts with vitamin C, it loses its pink colour. A solution of DCPIP is used which has enough DCPIP in 1cm³ of solution to react with 0.1mg of vitamin C.

Figure 25   *A titration method for finding the amount of vitamin C in a sample of cabbage.*

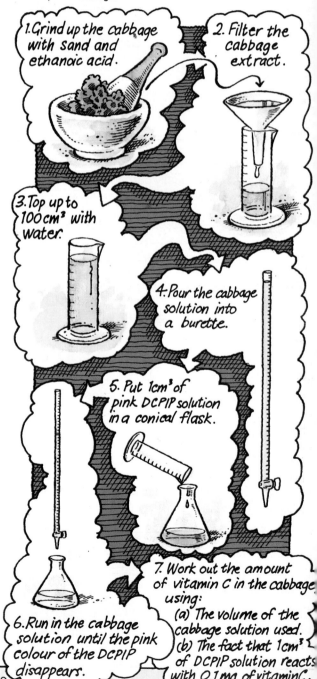

1. Grind up the cabbage with sand and ethanoic acid.

2. Filter the cabbage extract.

3. Top up to 100 cm³ with water.

4. Pour the cabbage solution into a burette.

5. Put 1cm³ of pink DCPIP solution in a conical flask.

6. Run in the cabbage solution until the pink colour of the DCPIP disappears.

7. Work out the amount of vitamin C in the cabbage using:
(a) The volume of the cabbage solution used.
(b) The fact that 1cm³ of DCPIP solution reacts with 0.1 mg of vitamin C.

Titrations are very important in chemistry. Chemists use titrations, like the one shown in figure 25, to find how much of one solution reacts with another solution.

Titrations are often used to measure the concentration of an acid. The acid is measured out, then titrated with an alkali. An **indicator** is used to show when the right amount of alkali has been added.

Titrations may be used:

◆ in the soft drinks industry – to check there isn't too much acid in a drink,
◆ in a pharmacy – for example, to check the amount of iron sulphate in iron tablets.

3   What are oils and fats?

Figure 26 ▶
Oils are liquids and fats, like the margarine in this photograph, are solids but their chemical structures are similar.

Liquid oils and solid fats are an important part of our diet. They are essential to the human body, although too much can be bad for your health.

When we talk about 'oils' in this section, we mean *edible* oils, like olive oil or corn oil. These are very different from the inedible oils, like motor oil, which are used to lubricate machinery.

Oils and fats have similar chemical structures. Their molecules contain long chains of carbon atoms with hydrogen atoms attached (figure 27). You can see from figure 27 that oils have more carbon-carbon double bonds than fats. Substances with double bonds are described as **unsaturated**. These double bonds help to give oils a lower melting point and this is why they are liquids.

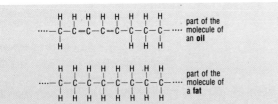

Figure 27   *The structures of oils and fats*

Chemists can change oils to fats by turning their double bonds to single bonds by adding hydrogen – they became **saturated**. Figure 28 shows what happens when a double bond is **hydrogenated**. This is an example of an **addition reaction**.

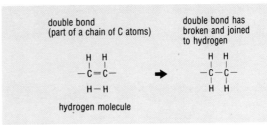

Figure 28   *Hydrogenation of a double bond*

This reaction does not work at room temperature. It needs a catalyst made of nickel and a temperature of 150 °C.

Hydrogenation is used to turn oils to fats. The oil is heated to 150 °C and a powdered nickel catalyst is added. Hydrogen is then bubbled through the oil. This changes double bonds in the oil to single bonds. The more hydrogen that is used, the more double bonds react (figure 29).

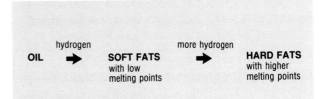

Figure 29   *Turning oil to fats*

What is margarine?

Margarine is a kind of artificial butter, which many people now prefer to butter. Like butter, it is an emulsion. Margarine contains droplets of water suspended in oil and fat. An emulsifier is used to make the oil and water mix.

Hydrogenation is an important part of making margarine. Figure 30 summarises the process.

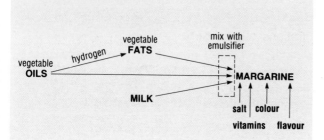

Figure 30 *The process for making margarines*

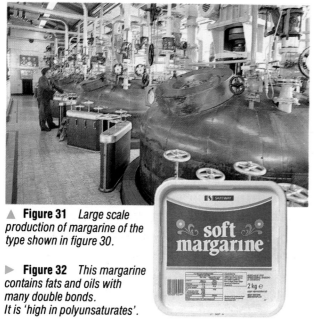

▲ **Figure 31** *Large scale production of margarine of the type shown in figure 30.*

▶ **Figure 32** *This margarine contains fats and oils with many double bonds. It is 'high in polyunsaturates'.*

Different kinds of margarines contain different amounts of oils and fats. Soft margarines contain more oil and softer fats. Some margarines are described as 'high in polyunsaturates'. This means they have lots of double bonds in their molecules. All margarines have flavourings and colourings added to them to make them taste and look like butter.

It is quite easy to test fats and oils to see how unsaturated they are. Bromine does an addition reaction with double bonds between carbon atoms, in the same way that hydrogen does. The more double bonds there are in a fat or oil, the more bromine it will react with. As the bromine reacts, it loses its red colour. So a simple way to compare the amount of unsaturation in different fats is to see how much bromine they can decolourise. The fat is shaken with a solution of bromine in water, called bromine water. The more bromine water the fat can decolourise, the more unsaturated the fat must be.

Doctors believe that unsaturated fats may be healthier than saturated fats. You can read more about this in page 107 earlier in this chapter.

Taking it further

Fats and oils are esters. They are made by combining an alcohol called glycerol with acids called **fatty acids**. Fatty acids contain long chains of carbon atoms.

A common saturated fat is glyceryl tristearate (figure 33). It contains glycerol combined with stearic acid.

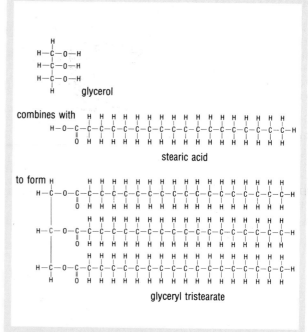

Figure 33 *Glyceryl tristearate has no double bonds between its carbon atoms. It is saturated.*

A common unsaturated fat is glyceryl trilinoleate. Its structure is shown in figure 34.

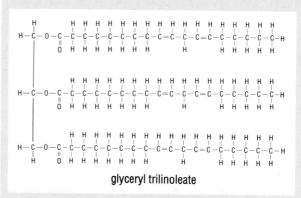

Figure 34 *Glyceryl trilinoleate has six double bonds between carbon atoms in its molecule. It is highly unsaturated.*

Figure 35   *The jug contains the main ingredients for French dressing - oil and vinegar (which is mostly water). They do not mix - the oil floats on the vinegar.*

Oil and water don't mix. They are **immiscible**, as you can see from the ingredients for French dressing in figure 35.

Oil and water don't mix because their molecules are very different from one another. Each prefers to stay with its own kind, rather than getting mixed up with the other (figure 36).

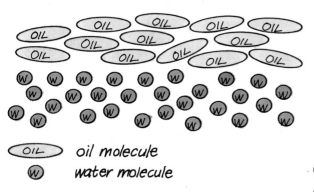

⬯ *oil molecule*

ⓦ *water molecule*

Figure 36   *Oil molecules and water molecules prefer to stick with their own kind, rather than get mixed up with each other.*

Emulsifiers – the go-betweens

An **emulsion** is a mixture of two liquids. In an emulsion the first liquid forms tiny droplets which are suspended in the second liquid.

Some emulsions consist of tiny droplets of oil suspended in water:
　　　　an oil-in-water emulsion.
Some emulsions consist of tiny droplets of water suspended in an oil:
　　　　a water-in-oil emulsion.

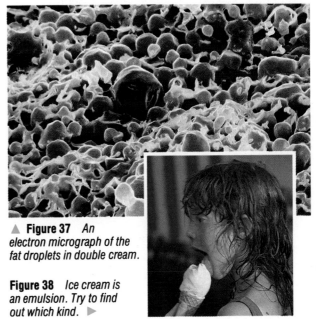

▲ **Figure 37**   *An electron micrograph of the fat droplets in double cream.*

Figure 38   *Ice cream is an emulsion. Try to find out which kind.* ▶

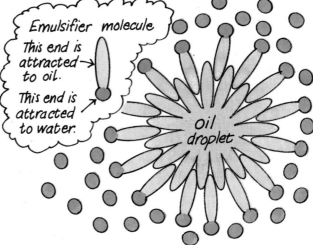

In cooking we often want to make oil and water mix because water makes the oil feel lighter in the mouth. You can make oil and water form a *temporary* emulsion just by shaking them together vigorously. This breaks up the liquids into tiny droplets. However, the liquids soon separate, as the droplets join together again. Oil prefers oil and water prefers water.

What is needed is a 'go-between' to bring the two liquids together. These 'go-betweens' are called **emulsifiers**.

Emulsifiers have molecules which are attracted to both oil molecules and water molecules. Figure 39 shows how emulsifiers act as 'go-betweens'.

Emulsifier molecule

This end is attracted → to oil.

This end is attracted ↗ to water.

oil droplet

Figure 39   *How an emulsifier helps oil and water mix*

Notice how similar this is to the working of a detergent, described on page 44. Indeed, detergents make very good emulsifiers.

Emulsions you eat

Many foods are emulsions. Table 4 lists some common food emulsions and their emulsifiers.

Some of the foods in the table are natural emulsions and some are made artificially. Artificial emulsions have emulsifiers specially added to them. You can see the emulsifiers listed among the ingredients on the packet. Emulsifiers have E numbers beginning with 3 or 4.

Table 4 *Some common food emulsions.*

Name of the food	Contains droplets	Droplets suspended	Emulsifier
Milk and cream	Oil	Water	Milk proteins
Butter	Water	Oil	Milk proteins
Mayonnaise (traditional)	Oil	Water	Egg yolk, which contains a natural emulsifier called lecithin
Salad cream	Oil	Water	Egg yolk, mustard, glyceryl monostearate (E471), etc.
Margarine	Water	Oil	Lecithin (E322), glyceryl monostearate (E471), etc.
Ice-cream	Oil	Water	Glyceryl monostearate (E471), proteins, etc.

Taking it further

Emulsifiers have molecules with an 'oil-loving' end and a 'water-loving' end. The oil-loving end usually contains long chains of carbon atoms, like oils do. The water-loving end usually carries an ionic charge or the O-H group. This is attracted to water's O-H groups.

For example, glyceryl monostearate (E471) is often used as an emulsifier. Its molecule is shown in figure 40.

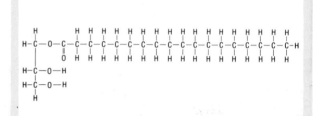

Figure 40

1 Which part of the glyceryl monostearate molecule is attracted to oil?
2 Which part is attracted to water?

Look back at the structure of the saturated fat glyceryl tristearate on page 112. You will see that glyceryl monostearate is quite similar, except it has only one carbon chain instead of three.

5 What is chromatography?

Artificial colourings are often used to make food look attractive. In recent years, scientists have found that some of the colourings that are commonly used may be dangerous to health. Some colourings have been banned (see page 106).

Public analysts have the job of making sure that food is safe to eat. From time to time they check the colourings used in foods. They sometimes use a process called **chromatography** to do this.

Figure 41 *Many fruit drinks now only use natural colourings, like carotene from carrots. Some drinks now use no colourings at all.* ▶

Figure 42 shows how chromatography could be used to check the colourings in an orange drink.

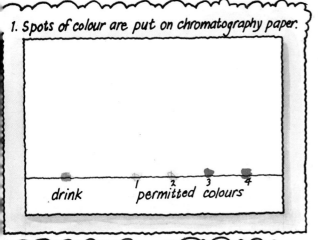

1. Spots of colour are put on chromatography paper.

drink   permitted colours

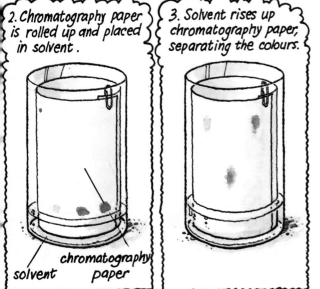

2. Chromatography paper is rolled up and placed in solvent.

3. Solvent rises up chromatography paper, separating the colours.

solvent   chromatography paper

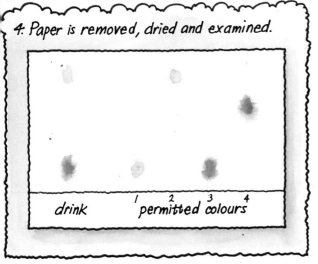

4. Paper is removed, dried and examined.

drink   permitted colours

Figure 42 *Using chromatography to check the colours in an orange drink.*

The chromatogram is 'run' by standing it in a suitable solvent. As the solvent rises up the paper, it carries the colours with it. Some colours dissolve in the solvent very well and because of this they are carried a long way up the paper by the solvent. Other colours do not dissolve so well and are attracted to the paper. These colours do not get carried so far up the paper.

The finished chromatogram is then examined by the public analyst. Look at the final picture in figure 42 You can see that the colour from the orange drink has been separated into a yellow colour and a red colour. The yellow colour must be the same as Permitted Colour number 2, because it has travelled the same distance as that colour. The red colour in the drink must be the same as Permitted Colour number 3. So the public analyst could conclude that the orange drink colour is made up of two permitted colours.

What are R_f values?

Sometimes it is possible to examine a chromatogram by eye', like the one in figure 42. You can judge by eye how far the different spots have moved. Sometimes, however, more accurate measurements have to be made to compare *how far* the different spots have moved.

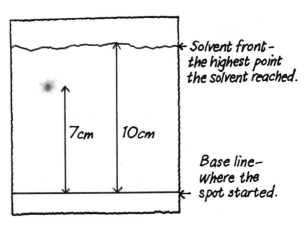

Solvent front – the highest point the solvent reached.

7cm   10cm

Base line – where the spot started.

$$R_f \text{ value} = \frac{\text{distance moved by coloured spot}}{\text{distance moved by solvent front}} = \frac{7cm}{10cm} = 0.7$$

Figure 43 *A chromatogram after removal from the solvent.*

Look at figure 43. The coloured spot has moved 7 cm up the chromatography paper compared with the 10 cm which the solvent has moved.

We say that this particular spot in figure 43 has an R_f value of 0.7, because it moves 0.7 of the distance moved by the solvent front. ('R_f' stands for 'relative to the front'.)

The R_f value of a particular colour is fixed, provided the solvent, paper type and temperature are fixed. This makes R_f values a useful way of comparing the colours separated by chromatography.

What other uses does chromatography have?

Chromatography can be used to separate colourless substances as well as coloured ones. After the chromatogram has run, the colourless substances are separated but cannot be seen. The chromatogram is 'developed' by spraying the paper with a special 'locating agent'. This makes the spots become coloured so they show up.

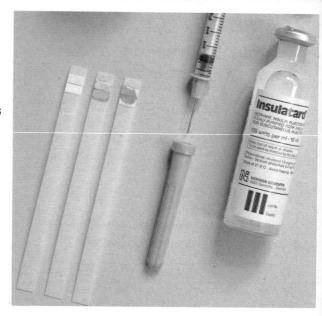

Figure 44 *The test strips in this photograph show the blood sugar level of a diabetic. The centre strip is normal. The right-hand strip shows the blood sugar is high and that treatment is required. (The left-hand strip is unused).* ▶

Taking it further

The flavours of foods come from a complex mixture of chemicals. Onion oil, for example, is extracted from onions and has a very strong onion flavour. Onion oil contains over 50 different chemicals. These 'flavour chemicals' evaporate from the onion and reach your nose as gases. Each contributes towards the flavour of onions.

The people who manufacture food flavourings need to be able to separate flavours into their different components. To do this, they use **gas chromatography** (gc). Figure 45 outlines how this works.

A stream of inert 'carrier' gas, for example, nitrogen or argon, carries the flavour sample through the column. The column contains a liquid, coated onto an inert solid.

Different components of the flavour dissolve in the liquid to different extents. The ones that dissolve *most* are held back in the column. The ones that dissolve *least* are quickly carried through by the gas. In this way, the different flavour components are separated. As the different components come separately out of the column, they are detected and recorded on a chart recorder.

This may seem very different from the paper chromatography you have used but the basic idea is the same. In both cases there is a moving phase (solvent in paper chromotography, gas in gc) and a stationary phase (paper in paper chromatography, liquid in gc). Between them, these two phases separate the components.

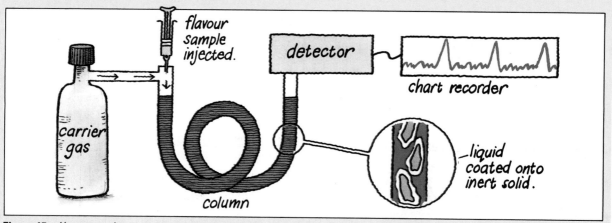

Figure 45 *How gc works*

Things to try out

Try these out at home. Remember – you should **never** taste food in a science laboratory.

1 *Tasting crisps* Can you tell the difference between crisp flavours? Get bags of at least four different flavour crisps. Plan a way of tasting the crisps without knowing which is which. How many flavours can you detect? Try it on a friend or a member of your family.

2 *Making sherbet* Sherbet is a mixture of sodium hydrogencarbonate, citric acid and sugar. When these powders dissolve in the water in your mouth, the sodium hydrogencarbonate and citric acid react. They fizz and give off carbon dioxide. You can buy citric acid and sodium hydrogencarbonate from a chemist's shop.
Recipe
Put the following ingredients into a bowl and mix very thoroughly:
9 teaspoonful icing sugar
2 teaspoonful citric acid
1 teaspoonful sodium hydrogencarbonate
Keep the sherbet in a sealed container.

Things to find out

3 Find out about the following different kinds of milk and cream. How do they differ from ordinary milk?

Homogenized milk; skimmed milk; semi-skimmed milk; single cream; double cream.

4 Find out about self-raising flour. What is in it that makes it different from plain flour?

5 Most flavourings are a mixture of many different substances. However, a few chemicals have a recognisable flavour of their own. Find out the flavour of each of the following:

benzaldehyde; pentyl ethanoate (also called amyl acetate); butyl butanoate; ethanal (also called acetaldehyde); 3-methylbutyl ethanoate (also called isoamyl acetate).

6 Make a survey of food additives in food in your home. Collect together about 20 or more different food items. Look at their ingredients lists, if they have them. Decide which ingredients are additives, and which are food. (Most additives, apart from flavourings, will have numbers. Table 2 on page 106 will help you crack the numbering code if necessary.)

a) Which foods contain colouring additives?
b) Which foods contain flavouring additives?
c) Which foods contain emulsifiers?
d) Which foods have no additives at all?
e) Which foods have no list of ingredients? Does this mean these foods have no additives in them?

Making decisions

7 *Should we use irradiation?* Irradiation is a quick and convenient new method for preserving food. The idea is to kill microbes in food by passing gamma rays through it. The gamma rays come from a radioactive material, usually cobalt-60. The food is prepared and packaged, then it is brought near to the radioactive material. The gamma rays go through the packaging and the food, killing microbes on the way.

Irradiation can be used to preserve fruit, vegetables and many other foods. The food does not become radioactive itself, provided the strength of the gamma rays is not too high. But if the strength of the rays is too *low*, the microbes do not get killed.

Like other food preservation methods, irradiation may change the taste of foods. The amount of change depends on the food involved. The radiation breaks down some of the chemicals in the food, and this produces new chemicals.

Irradiation of food is permitted in some countries of the world but banned in others.

Suppose you are a member of a government committee investigating food irradiation. You have to decide whether irradiation should be permitted in this country.

a) What information would you need before you could make your decision?
b) What experiments or tests would have to be done to provide the information you need?

Points to discuss

8 Should we eat more fresh food and less processed food? Would this be easy to do?

9 Is 'junk food' bad for you? Are some kinds of junk food worse than others?

10 Do food manufacturers 'push' junk foods on young people, and stop them eating healthier foods?

11 Do food colourings and additives mislead us, by making poor quality food look good?

12 Are some food additives more useful than others? If so, which?

13 Nutritionists recommend that practically everyone should cut down the amount of fat they eat. Would you find this easy to do?

14 Raw vegetables contain more vitamin C than cooked vegetables . Why don't we eat all our vegetables raw?

15 Is it justified to test food additives on animals to see if they are safe?

Questions to try

16 Fill in the missing words in the following passage. Each word is used only once.

The missing words are:
carbon, colouring, emulsifier, emulsion, flavouring, hydrogen, hydrogenation, saturated, unsaturated.

Fats and oils have similar chemical structures. Their molecules contain long chains of .... (a) .... atoms with .... (b) .... atoms attached to them.

Some fats and oils have double bonds between some of the atoms in the chains. They are called .... (c) .... fats. Fats with no double bonds are called .... (d) .... fats. You can turn the first kind to the second kind by a process called .... (e) ....

Margarine is made by blending fats and oil with milk. Margarine contains tiny droplets of water suspended in oil and fat. This is called an .... (f) .... To make the water and oil mix, it is necessary to add an .... (g) .... Margarine also contains .... (h) .... and .... (i) .... to make it more like butter.

17 a) What is baking powder? What job does it do?
   b) Sponge cakes taste light and fluffy because they contain lots of tiny bubbles. What gas do these bubbles contain?
   c) You are baking a cake. You need sodium hydrogencarbonate to make it rise. You discover there is a world shortage of sodium hydrogencarbonate. Which of the following substances would make the *best* replacement for sodium hydrogencarbonate?
      sodium carbonate, potassium carbonate, potassium hydrogencarbonate, lead carbonate, copper carbonate.

18 Suggest explanations for each of the following:
   a) You want to make a fruit salad using apples. The recipe advises you to sprinkle the apple with lemon juice immediately after cutting it up.
   b) Raw cabbage is more nutritious than cooked cabbage.
   c) Potato crisps are packed in bags containing nitrogen instead of air.
   d) You can buy hard margarine and soft margarine, but you can't buy hard and soft butter.

19 You are eating a green ice lolly. You reckon the green colour is a mixture of two dyes – blue and yellow. Your friend reckons it is a single green dye, not a mixture. Describe how you would find out who is right.

20 Michael wanted to find the amount of vitamin C in a sample of fresh orange juice.

He took $5cm^3$ of orange juice and added some ethanoic acid to make it acidic. He diluted the solution with distilled water, so that the final volume was $100cm^3$. He put some of this diluted solution in a burette.

Then he put $1cm^3$ of pink DCPIP solution in a conical flask. He ran in the diluted orange solution from the burette, until the pink colour of the DCPIP disappeared.

He found that $4cm^3$ or orange solution was needed.

Michael's teacher told him that $1cm^3$ of the DCPIP solution reacted with 0.1 mg of vitamin C.

   a) How much vitamin C was there in $4cm^3$ of *diluted* orange juice?
   b) So how much vitamin C was in $100cm^3$ of *diluted* orange juice?
   c) How much vitamin C was in the original $5cm^3$ of *undiluted* orange juice?
   d) The average daily requirement of vitamin C for a boy of Michael's age is 25 mg. How much orange juice would he need to drink to get his daily requirement?
   e) Michael wanted to repeat the experiment using blackcurrant juice instead of orange juice. Why might this have been tricky?

Introducing minerals

Figure 2 *Copper pyrites mineral is the green part of this rock. Copper is extracted from the copper ore but the ore usually contains less than 1% copper so a large amount of rock must be processed to produce useable amounts of copper.* ▼

Figure 1 *The rock which is dug from this massive copper mine in Australia contains the mineral copper pyrites ($CuFeS_2$).*

The radius of the earth is 6400 km. The deepest mine, which is a gold mine, is about 3.5 km. All of the resources which can be converted into the materials and objects which you use every day have to come from the thin crust of the earth or the sea or the atmosphere. If the earth was the size of an onion, then the crust would only be as thick as the skin!

In this chapter you will see how

◆ some of the important minerals are mined or quarried from the ground and how the reserves of these minerals differ,
◆ the useful parts of these minerals are extracted from the less useful parts,
◆ we depend on the materials made from these minerals and how the mining and extraction processes affect the environment.

Figure 3 *Sand contains the mineral silica (SiO_2). Silica is used to make glass. White sand is almost pure silica. What do you think makes some sand brown?*

1   How do minerals differ?

The pie chart in figure 4 shows the composition of the earth's crust. Some elements are much more abundant than others. Not all rocks containing a particular element are suitable for making that element. Sometimes an element is bonded too strongly to other elements in the rock. Sometimes there is so little of the element in the rock that it is not worthwhile extracting it.

A mineral is the useful part of a rock. Some rocks, such as salt and limestone are almost pure mineral.

Sometimes the only source of an element is a rock which has a small percentage of the element in it.

Sometimes a mineral is used in the form in which it is dug out of the ground. An element is not extracted from it. These minerals are called **mundane minerals**.

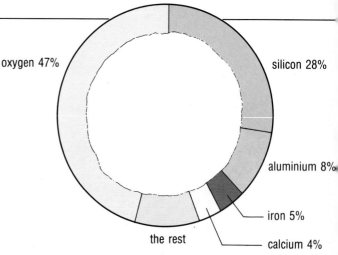

oxygen 47%

silicon 28%

aluminium 8%

iron 5%

calcium 4%

the rest

Figure 4   *The composition of the earth's crust*

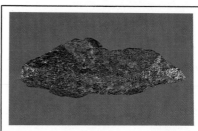

Less than 1% of copper iron sulphide ($CuFeS_2$).

Used as a source of copper metal.

Copper costs about £1500 per tonne.

Figure 6   *Copper pyrites*

It is about 75% aluminium oxide (Al_2O_3), 25% iron oxide (Fe_2O_3).

Used as a source of aluminium metal. Aluminium costs about £1200 per tonne.

Figure 5   *Bauxite*

0.0001% of uncombined gold in quartz (SiO_2).

Used as a source of gold. Gold costs about £10 000 000 per tonne.

Figure 7   *Gold-bearing quartz*

Typically about 15% lead sulphide (PbS).

Used as a source of lead. Lead costs about £330 per tonne.

Figure 8   *Galena*

Almost 100% calcium carbonate ($CaCO_3$).

Used as an ingredient in cement and glass.
Used for building and road making.

Figure 9 *Limestone*

Almost 100% sodium chloride (NaCl).

Mostly used to make chlorine and sodium hydroxide and sodium carbonate.

Figure 11 *Rock salt*

Often over 85% iron oxide (Fe_2O_3).

Used as a source of iron. Iron costs about £80 per tonne.

Figure 10 *Haematite*

Use the information next to figures 5 – 11 to answer these questions.

1 Which of the minerals is a mundane mineral?
2 Predict which minerals will produce the most solid waste material when used to extract a metal. Place the minerals in order of the amount of waste they will produce.
3 Which mineral is used as a source of an important non-metal?
4 Copper is more expensive than aluminium. Suggest a reason why.
5 Aluminium is more expensive than iron. Yet aluminium is more abundant in the earth's crust than iron, so you might expect it to be cheaper. Explain why it is not.

2 Saving glass

Glass is one of the most useful modern materials. It is strong, transparent, unreactive and easy to clean. And it is very cheap to produce. Figure 12 summarizes the way that glass is made.

The minerals used for making glass – sand, limestone and salt (used to make the soda ash) – are all very abundant. In fact, the composition of glass is similar to the composition of the earth's crust. So there is no shortage of raw materials and this is why glass is so cheap. The main cost in producing glass is not the cost of the raw materials but of the energy needed to melt them down together.

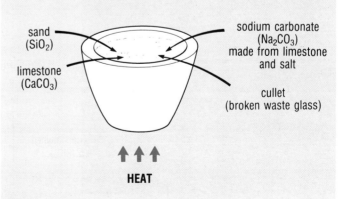

sand (SiO_2)

limestone ($CaCO_3$)

sodium carbonate (Na_2CO_3) made from limestone and salt

cullet (broken waste glass)

HEAT

Figure 12 *Making glass. This is the method used to make the commonest type of glass.*

Why save glass?
Even though it is cheap, there are good reasons for saving glass.

◆ The raw materials for making glass may be abundant but they still have to be dug from the ground. Saving glass means we need fewer quarries.
◆ Saving glass saves fossil fuels.
◆ Waste glass causes a litter problem (figure 13). Because it is very unreactive, glass does not get broken down by bacteria. Waste glass stays around for centuries. Broken glass can cause injuries to people and animals. Empty glass containers can act as traps for unsuspecting mammals. (An empty milk bottle found in Essex contained the skeletons of 28 small animals!) Glass litter can even cause fires.

1 Explain why saving glass helps save fossil fuels.
2 Suggest how glass litter may help cause fires.
3 Make a list comparing the litter problems caused by glass and by plastic. Which causes the worst problem?

◀ **Figure 13**
Glass litter like this will never biodegrade and must be physically removed to avoid damage to people and animals.

Figure 14 *Milk bottles are strong and because empty bottles are collected by the delivery person they can be cleaned and reused many times.*

How can we save glass?
There are two ways to save glass: reusing and recycling.

Reusing involves using glass containers several times. This is possible because glass is so strong and hard-wearing. Milk bottles are a good example of reusing glass. The average milk bottle makes 25 trips to the doorstep in its lifetime. The bottles are returned to the dairy, cleaned (figure 14) and refilled. All this uses energy but less energy than is needed to make a new bottle.

But most British bottles and jars are not designed to be refillable. The problem is, people do not want the trouble of returning glass containers, unless it's as easy as putting a milk bottle on the doorstep. They would rather throw away the container and buy a new one.

Recycling involves collecting waste glass and using it to make new glass. When waste glass is used in this way it is called **cullet**. The broken cullet is added to the ingredients used to make glass and melted down with them (see figure 12).

Less energy is needed to melt down cullet than to melt the ingredients of glass. So recycling glass like this saves energy as well as raw materials. But the saving in money is quite small. Unless the recycling can be done cheaply it will not pay for itself.

The problem with recycling is that you have to collect the waste glass and transport it to the glass works. The glass must be separated from other waste and sorted into colours. All this needs energy.

4 Think about reusing and recycling as ways of saving glass. Which do you think is better? Why?
5 You should never take milk bottles to the Bottle Bank. Why not?
6 You should always remove metal caps from bottles before putting them in the Bottle Bank. Why do you think this is? Why do paper labels not matter?
7 Where is your nearest bottle bank? Why do you think it was decided to put it there? Do you use it?
8 Apart from glass, what other forms of household waste can be easily recycled?

Gemstones have always fascinated people. They have been searched for, fought over, stolen and coveted for thousands of years.

Most gemstones are minerals. But what makes them different from other mineral substances is that they are very rare. The basic mineral they contain may be common. Opals are basically made of silicon oxide, the same mineral as in sand. But in a gemstone like opal the basic mineral is coloured and shaped in a particularly beautiful way.

Different kinds of gems

Table 1 gives details of some important gemstones. You can see that each gem contains a basic mineral, which makes up most of the gem. If the gem is coloured, the colour comes from traces of substances which are present as impurities.

◄ **Figure 15**
The Imperial State crown is covered in magnificent gemstones. Many of them have interesting histories. Find out what you can about them.

Table 1 *Important types of gemstones*

Type	Appearance	Basic mineral	Coloured by	Comments
diamond	clear	carbon (diamond)	—	The most important gemstone. Very hard.
amethyst	purple	quartz (SiO_2)	iron compounds	
sapphire	blue	corundum (Al_2O_3)	titanium and iron compounds	Can also be other colours – pink, yellow, etc.
emerald	green	beryl ($Be_3Al_2Si_6O_{18}$)	chromium oxide (Cr_2O_3)	The most expensive gemstone.
topaz	yellow	aluminium fluorosilicate ($Al_2F_2SiO_4$)	iron oxide (Fe_2O_3)	Many shades are possible, including brown.
ruby	red	corundum (Al_2O_3)	chromium compounds	

Most gemstones have definite crystalline shapes. This is because the atoms in them are arranged in a definite, regular way.

When the gem is dug out of the ground, it is roughly shaped and covered in dirt. But a skilled jeweller can cut and polish the gemstone so that its natural crystalline shape shows up (figure 16).

The surfaces of the crystal reflect light, making it sparkle. The crystal can also act like a prism, splitting up white light into colours. This gives the gemstone 'fire' – it flashes with colour.

How are gemstones formed?

Most gems were formed from molten rock, at very high temperatures and pressures. As the molten rock cooled, the gemstones crystallized out. Most of the gems formed like this were very small but a few grew to large sizes. These larger stones are the precious ones.

Many people have tried to make gemstones artificially. They mix together the substances which make up the gem. These are then heated to high temperatures and often put under high pressures. The trouble is, to make decent-sized gems, you need to do this for a long time. This costs a lot of money – often more than the gemstone would be worth.

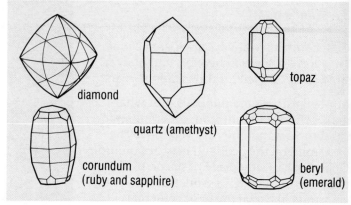

Figure 16 *The cut crystal shapes of some important gemstones*

1 Look at table 1. Write down the symbols of the elements that make up the *basic minerals* in the gems in the table. Now look at the Periodic Table on page 12. Find the positions of these basic mineral elements in the table. What do you notice?

2 Look again at table 1. Write down the symbols of the elements that *give colour* to the gems in the table. Now look again at the Periodic Table on page 12. Find the positions of these colouring elements in the table. What do you notice this time?

3 A particular type of gem can come in many different shades of *colour*. Yet the *shape* of a particular type of gem crystal is always the same. Suggest a reason why.

4 It is important that gemstones are very hard. Why?

5 Some of the gemstones in table 1 have uses apart from in jewellery. Give some examples.

4 *Aggregate – mundane but useful*

What is aggregate?

Aggregate includes sand, gravel and crushed stone (figure 17). Aggregate is used whenever things are being built – like houses, offices, roads and airports. Aggregate is an example of a mundane mineral: a mineral that is cheap and plentiful, and used in the form in which it is dug up.

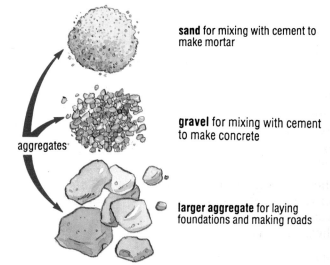

sand for mixing with cement to make mortar

gravel for mixing with cement to make concrete

larger aggregate for laying foundations and making roads

Figure 17 *Aggregates and their uses*

Figure 18 *Newly dug gravel is washed and sieved to produce different grades of aggregate before it is removed by lorry for use in the surrounding area.*

Huge quantities of aggregate are used in Britain each year.

◆ To build an average 3-bedroom house needs 50 tonnes of aggregate.
◆ To build 1 km of motorway needs 65 000 tonnes of aggregate.
◆ In one year, Britain uses about 200 000 000 tonnes of aggregate.

Where does all the aggregate come from?

Fortunately aggregate is very plentiful. Aggregate is mostly made of silicon oxide (SiO_2).

There are many places where you can dig up aggregate. This is just as well, because it would not be worth transporting such a cheap mineral long distances.

Table 2 compares the costs of buying and transporting a) gravel, b) bricks and c) copper piping.

Table 2 *Production and transport costs*

	Basic purchase cost/tonne	Transport cost/tonne	
		per 10 km	per 100 km
Gravel	£6	£2.50	£6.50
Bricks	£110	£3.50	£9.50
Copper piping	£3000	£6	£10.50

Use the figures in table 2 to answer these questions.

1 Suppose a builder needs a) 1000 tonnes of gravel, b) 100 tonnes of bricks and c) 1 tonne of copper piping, how much would each cost if they had to be brought
  i) from 10 km away,
  ii) from 100 km away?
2 If you were a builder, would you be concerned if your supplies of *gravel* were 100 km away? What if your supplies of *copper piping* were 100 km away?

Digging up aggregate makes big holes

Because of transport costs, it is necessary for aggregate to be dug up locally. There are few parts of Britain that are more than 40 km from a sand or gravel quarry.

Deposits of aggregate usually lie near to the surface of the ground. Most deposits were laid down by rivers or seas. This means that they are in valleys or other low ground. So aggregate deposits are usually easy to get at. But towns and other settlements are often built on low ground too. This means that the quarries are often near built-up areas.

The amounts of aggregate that must be dug up are very large – about 80 million cubic metres per year in Britain. So quarrying aggregate leaves some very big holes. The holes can be ugly and dangerous (figure 19).

3 Where is the nearest sand or gravel pit to your home?
4 Sand and gravel pits are often full of water. Why?
5 What use can be made of disused sand and gravel pits?
6 Suppose you are Minister for the Environment. What laws would you try to make to reduce the damage done to the environment by digging up aggregate?

Figure 19 *Gravel pits leave large holes when excavation is completed. What do you think these holes could be used for?*

In brief
Minerals

1 Minerals are materials found in the earth's crust and from which useful substances can be extracted. This means that coal and oil can also be thought of as minerals.

2 A mineral is the useful part of a rock. Rocks from which metals can be extracted are called **ores**. A high grade ore is one which contains a lot of the mineral.

the form in which they are dug up. Minerals used in this way are called **mundane minerals**. For example, sand is fed directly into the furnace when it is used to make glass.

3 Once a mineral has been taken out of the ground and used it will not be replaced. Mineral resources are finite which means there could come a point when a particular mineral is all used up. One way to make minerals last longer is to recycle metals and other mineral products.

4 The decision of a company to open or close a mine must be based on a consideration of:

Scientific evidence	– What grade is the ore?
Technical evidence	– How easy will it be to extract the ore?
Market predictions	– What is the likely demand and what can be charged for the product?
Environmental and social factors	– What effects will it have on the environment and the local community?

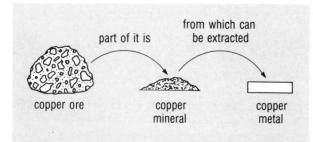

part of it is → from which can be extracted →

copper ore   copper mineral   copper metal

Figure 20

Minerals such as limestone and sand are not used as sources of elements. They are used in

5 Some of the benefits and drawbacks of mining are summarised below:

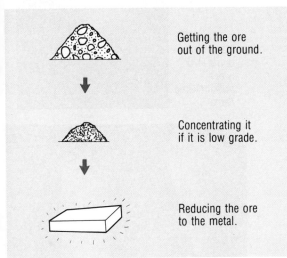

Drawbacks	Benefits
Mining and quarrying can affect the environment, particularly the ecology of the area.	Additional employment
	Produces useful materials for people to use
The extraction process may produce waste material which has to be disposed of.	Creates wealth for the company and the country

6 Extracting a metal involves the processes in figure 21.

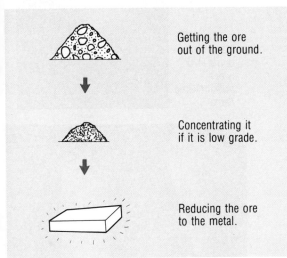

Getting the ore out of the ground.

Concentrating it if it is low grade.

Reducing the ore to the metal.

Figure 21

Lead, zinc and copper occur as sulphides, which must first be roasted to convert them to oxides. Iron and aluminium occur as oxides and gold occurs as the element uncombined with anything. Some of the very reactive metals, such as sodium, magnesium and aluminium, are extracted by electrolysis (see the chapter on Making and Using Electricity).

7 When a sulphide ore is roasted in air sulphur dioxide is formed. This gas can be mixed with air (oxygen) and passed over a catalyst to form sulphur trioxide. This can then be converted into sulphuric acid.

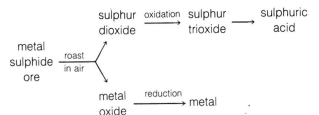

8 A catalyst changes the speed of a chemical reaction without being used up itself. This can be investigated by adding manganese(IV) oxide powder to a solution of hydrogen peroxide (figure 22).

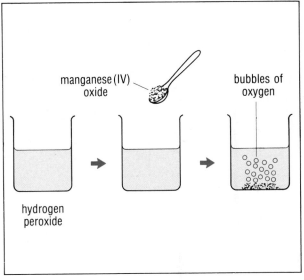

Figure 22

9 More limestone is extracted from the ground than any other mineral. Limestone is called a mundane mineral because for most of its uses it can be used in the form it comes out of the ground

For some purposes it is first heated which converts it from limestone (calcium carbonate) into quicklime (calcium oxide).

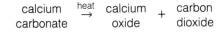

Most other carbonates, except those of sodium and potassium, decompose in a similar way to give the metal oxide and carbon dioxide.

10 Relative atomic masses of elements are used in calculations rather than actual masses because atoms are so very small and light. Relative masses are their masses compared to a standard. They can be used to make predictions about the mass of substances involved in reactions.

11 The Periodic Table is a way of arranging the elements and displaying them which helps you to remember the similarities and differences between them. For example, the table collects together elements which are similar into vertical columns called **groups**.

1 What do methods of extracting metals have in common?

Look at the flow diagram (figure 23). It shows how the techniques used to extract metals depend on the answers to several important questions. You could use figure 23 to study the methods used to extract sodium, magnesium, aluminium, iron, lead, zinc and copper. If you wanted to include unreactive metals such as gold you would need to extend the diagram.

▼ **Figure 23**

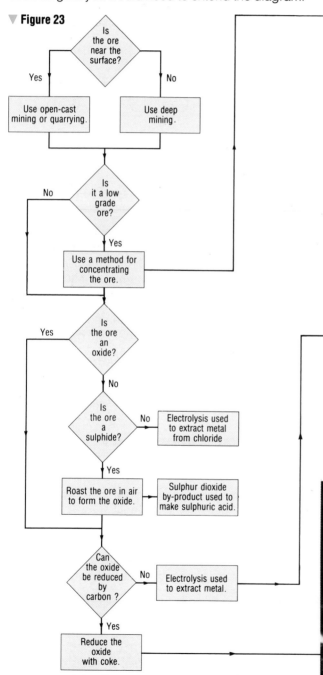

Figure 24 *A settling pond for bauxite at an aluminium refinery*

► **Figure 25** *Magnesium burns in air with a brilliant white flame*

Figure 26 *An ingot of pure aluminium*

Figure 27 *Tapping molten iron from a blast furnace for further processing into steel of different grades* ▶

2 How can low grade ores be made more concentrated?

Whatever method is used to concentrate the ore it will work better if the ore is a powder rather than big lumps. So the first stage is to crush the ore. The crushing and grinding uses energy and so adds to the cost of the final product.

The methods used to concentrate the ore depend on differences in physical or chemical properties between the mineral and the waste material. Here are two examples of methods of concentrating ores.

Copper ore

Copper ores are very low grade. This means the percentage of mineral present is small. The mineral is separated from the waste material by **froth flotation**.

In this process the crushed ore is churned up with water and oils. A froth forms on the top. Most of the mineral sticks to the froth and the waste material sinks.

Aluminium ore

Bauxite is the ore of aluminium. It is concentrated by:

first dissolving the aluminium oxide out of the ore with sodium hydroxide solution,

then separating the solution from the insoluble material

and finally crystallising the purified aluminium oxide out of the solution.

This is called **chemical leaching**. A similar method is used in modern gold mining but a different chemical is used to dissolve the gold.

Any method of concentrating an ore leaves the problem of how to dispose of the waste material. Often there are huge quantities of waste. For example, in the production of one kilogram of pure copper, 500 kg of waste is produced.

Figure 28 *Froth flotation (left) is used to separate the copper mineral from the copper-bearing ore in tanks like these (right).*

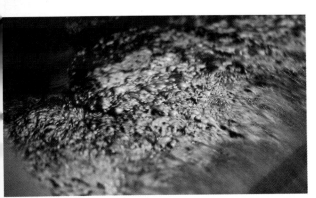

Taking it further

Metal oxides react with acids to form salts and water. For example.

sulphuric acid	+	copper oxide	→	copper sulphate	+	water
H_2SO_4	+	CuO	→	$CuSO_4$	+	H_2O

In this reaction the metal oxide is acting as a **base** and reactions of this type can be summarised as

acid + base → salt + water

Some metal oxides will react with strong alkalis as well as with acids. This type of reaction is used to concentrate aluminium ore.

Bauxite is a mixture of aluminium oxide and iron oxide. When it is treated with sodium hydroxide solution, which is a strong alkali, the aluminium oxide dissolves but the iron oxide does not. In this reaction the aluminium oxide is acting as an acidic oxide.

aluminium oxide	+	sodium hydroxide	→	sodium aluminate	+	water
Al_2O_3	+	$2NaOH$	→	$2NaAlO_2$	+	H_2O

Once the insoluble iron oxide has been removed by filtering, the sodium aluminate is converted back to aluminium oxide.

Aluminium oxide has some basic and acidic properties. Metal oxides which react with both acids and alkalis are called **amphoteric oxides**.

3 How is sulphuric acid made?

Over 2.5 million tonnes of sulphuric acid are made in the UK each year. As you can see from figure 29 it is used to make a variety of important substances.

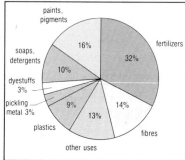

Figure 29 ▶

Sulphuric acid is made from sulphur dioxide which can come from three different sources.

a) Much sulphur is obtained by extracting it from crude oil. This has the added advantage that if the sulphur is removed from the oil there is less chance of sulphur dioxide air pollution when the oil is burnt.

Sulphur dioxide is made by burning this sulphur.

$$\text{sulphur} + \text{oxygen} \rightarrow \text{sulphur dioxide}$$
$$\text{S} + \text{O}_2 \rightarrow \text{SO}_2$$

b) In some countries underground deposits of pure sulphur have been found. It is not combined with any other element so all that has to be done is to extract it from the ground and burn it to form sulphur dioxide.

c) When sulphide minerals are roasted in air to convert them to the metal oxide, sulphur dioxide is also formed. For example,

$$\text{zinc} + \text{oxygen} \rightarrow \text{zinc} + \text{sulphur}$$
$$\text{sulphide} \qquad\qquad \text{oxide} \quad \text{dioxide}$$
$$2\text{ZnS} + 3\text{O}_2 \rightarrow 2\text{ZnO} + 2\text{SO}_2$$

Having made sulphur dioxide, the next step in the process is to convert it to sulphur trioxide. This is done by mixing the sulphur dioxide with air and passing it over a catalyst called vanadium(V) oxide at 450°C. The sulphur dioxide is converted to sulphur trioxide.

$$\text{sulphur} + \text{oxygen} \rightarrow \text{sulphur}$$
$$\text{dioxide} \qquad\qquad \text{trioxide}$$
$$2\text{SO}_2 + \text{O}_2 \rightarrow 2\text{SO}_3$$

This stage is called the **Contact Process** because when the gases come into *contact* with the catalyst, the reaction between them is speeded up.

The sulphur trioxide is then reacted with water to form sulphuric acid.

$$\text{sulphur trioxide} + \text{water} \rightarrow \text{sulphuric acid}$$
$$\text{SO}_3 + \text{H}_2\text{O} \rightarrow \text{H}_2\text{SO}_4$$

4 Why are catalysts used?

In industry catalysts are used to speed up reactions and allow them to happen at lower temperatures. This saves energy and so makes the products cheaper.

Catalysts are used to make:

◆ sulphuric acid (and hence detergents, paints),
◆ ammonia (and hence fertilizers),
◆ nitric acid (also for fertilizers),
◆ ethene (and hence plastics),
◆ in the food industry to make margarine.

Because of this catalysts have a big impact on our lives and chemists are always searching for new catalysts.

The efficient working of our bodies also depends on catalysts. These are enzymes and are called **biological catalysts**. The chapter on Fighting Disease discusses them in more detail.

Catalysts work by the reacting substances becoming temporarily attached to the catalyst (figure 30a and b).

While they are there they react together more easily (figure 30c).

Then the products are released leaving the catalyst unchanged (figure 30d).

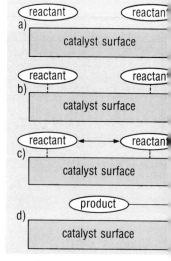

Figure 30 ▶
How a catalyst works

When solid catalysts are used, the reaction must occur at the surface of the catalyst. The catalyst will be more efficient if it is made in a form which has a large surface area (figure 31).

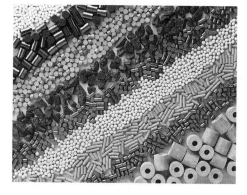

◀ **Figure 31**
Small separate catalysts provide large surface areas for the reactions they are used in.

5 How is iron obtained from iron ore?

There are large reserves of iron ore in many countries. Haematite (iron oxide, Fe_2O_3) is one of the main ores. The iron oxide has to be reacted with something which has a stronger attraction for the oxygen than iron does. The cheapest substance which will do this is carbon in the form of coke. The process of removing oxygen is called **reduction**.

Haematite is a high grade ore and does not need to be concentrated. It can be crushed and fed directly into the furnace where it is reduced to iron.

Iron ore, coke and limestone are added at the top of the furnace (figure 32). The hot air burns some of the

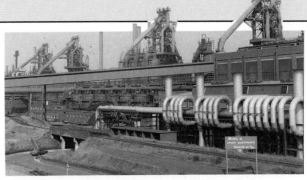

Figure 33 *These furnaces get their name from the hot air which is blasted in to them. They are over 60 metres high.*

coke and forms carbon monoxide. The reaction is exothermic and gives out a lot of heat.

$$\text{carbon} + \text{oxygen} \rightarrow \text{carbon monoxide}$$
$$2C + O_2 \rightarrow 2CO$$

The iron oxide is reduced by the carbon monoxide.

$$\text{iron oxide} + \text{carbon monoxide} \rightarrow \text{iron} + \text{carbon dioxide}$$
$$Fe_2O_3 + 3CO \rightarrow 2Fe + 3CO_2$$

The temperature in the furnace becomes so high that the iron which is formed is melted. It sinks to the bottom of the furnace.

The impurities in the ore react with the limestone and form a liquid 'slag' which floats on top of the iron.

$$\text{limestone} + \text{impurities} \rightarrow \text{'slag'}$$

The furnace is run continuously. Raw materials are fed into the top and the iron and slag are run off at the bottom.

The iron is made into steel by refining it and adjusting the impurities in it so that they give the steel particular properties.

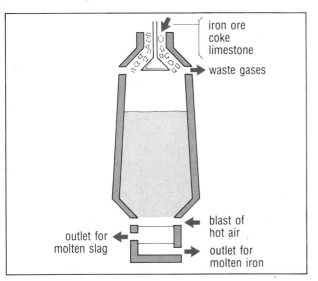

iron ore
coke
limestone

waste gases

blast of hot air

outlet for molten slag

outlet for molten iron

Figure 32 *A blast furnace*

Taking it further

The iron made in a blast furnace is too brittle for many uses. It is called **pig iron** or **cast iron**. Most of the iron is converted to steel. This is done by blowing oxygen into the molten metal. The impurities in the iron, mostly carbon, are burned to form gases which escape. Then carefully calculated amounts of impurities are added back to the iron to give it the special properties needed. So steel is just iron with particular impurities present.

Stainless steel, which does not rust and is used for cutlery, is 70% iron, 20% chromium and 10% nickel (figure 34).

The '**mild steel**' used to build motor-car bodies consists of 99.8% iron and 0.2% carbon (figure 35).

▼ **Figure 34** *Stainless steel cutlery is made from more highly refined steel than is used for car bodies.*

Figure 35 *Sheets of mild steel are pressed in the body works of car factories to produce the various panels used in the construction of a car body.*

6 Why is so much limestone quarried?

Limestone is almost pure calcium carbonate. It is a sedimentary rock. It was formed millions of years ago from the skeletal remains of creatures living in the sea. Some deposits of limestone were subjected to heat and pressure and changed to the much harder metamorphic form called **marble**.

Limestone is often used unchanged. But sometimes it is converted to quicklime (calcium oxide) or slaked lime (calcium hydroxide) before it is used.

▲ **Figure 37** The Taj Mahal at Agra in India is built of marble into which semi-precious stones are inlaid.

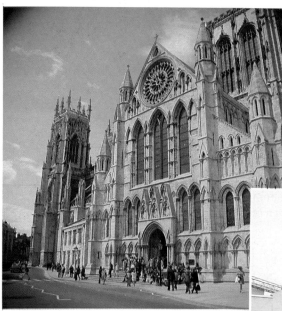

Figure 36 Magnesium limestone was used to build York Minster. Magnesium limestone contains both calcium carbonate and magnesium carbonate.

◄ **Figure 38** Some limestone is turned into quicklime (calcium oxide) by heating it in large rotating cylindrical kilns.

Figure 39 Crushed limestone is heated with soda ash and sand to form glass.

Figure 40 This quarry produces limestone which is crushed for use in road making.

Sometimes it is necessary to take a course of **antibiotics**. Some antibiotics, called tetracyclines, attack the bacteria and they also appear to reduce the production of sebum. However, it is important that courses of antibiotics should be used only when necessary as overuse can encourage the development of resistant strains of bacteria. This is one of the most important issues facing chemists trying to fight disease. It is also mentioned in the chapter on Fighting Disease.

Keeping your skin clean

It is important to keep your skin clean as it will reduce the chances of the hair follicles becoming blocked by dead skin and dirt. But the production of too much sebum deep down in the skin is the fundamental cause of the problem. However clean you keep your skin, blockage of the sebacious gland can still cause a spot to develop.

The production of sebum is influenced by changes in your hormones which occur during adolescence. This is why acne is particularly common among young people. However, the fact that the condition will improve after adolescence does not mean that it should not be treated.

1 If you were asked to advise a young person who was suffering from spots, what would you want to say to her or him? Using the information in this section make a list of facts you would wish to explain to her or him and place them in order of priority.

2 Acids, alkalis and oxidising agents are names for particular groups of chemicals.

◆ All acids have some properties in common with each other.
◆ All alkalis have some properties in common.
◆ All oxidising agents have some properties in common.

Find where these names are mentioned in this section and write a sentence about each which explains why they are mentioned.

2 Hair and the things we do to it

How do you treat *your* hair? We all have to keep washing our hair but some people add conditioner, some dye it, some perm it, some bleach it, some use gel and some use hair-setting spray. All these operations use chemicals and it is worth considering how the chemicals work.

The sebacious gland in each hair follicle produces an oily liquid called sebum (see figure 6). This liquid, besides making the hair smooth and shiny, protects it by controlling loss of water from the hair and limiting the growth of bacteria and fungi. However, the sebum is sticky and attracts dirt and so everyone has to wash their hair fairly frequently.

Shampoos contain detergent which is the part of the shampoo responsible for cleaning the hair. The detergent breaks up the sebum into tiny droplets which can then be washed away by water. Ideally the shampoo should still leave a thin coating of sebum on the hair to protect it and keep it shiny.

Figure 9 *The things we do to our hair!*

Figure 10 *What would you expect the differences to be between these different shampoos?* ▶

Before considering any other differences in shampoos let us look more closely at what hair is made of.

Each hair is made of a large number of protein fibres. The structures of proteins are complicated. They consist of lots of small molecules joined together. Each of these small molecules is an **amino acid**. One protein differs from another in the number of different amino acids it contains and the sequence in which they are arranged (see page 53).

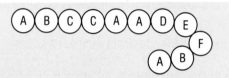

Figure 11 *A diagrammatic representation of part of a protein molecule which contains six different amino acids*

As the name amino acid implies, each one contains a part which is similar to ammonia (an alkali) and a part which is acidic. The amino parts are slightly positively charged and the acid parts are slightly negative. The positive parts of one protein fibre are attracted to the negative parts of the next protein fibre. (The meaning of the word fibre is discussed on page 41.)

Whenever you put liquids on your hair you tend to affect the crosslinks between the protein fibres in your hair (figure 12). The simplest example of this is when you wet your hair with water. Some of the links are broken by the wetting and then others reform as your hair dries.

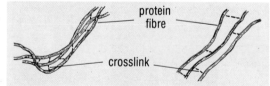

Figure 12 *The protein fibres in hair are held together by crosslinks.*

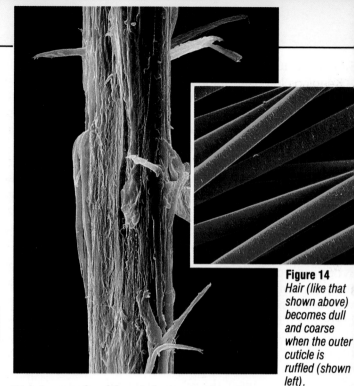

Figure 14
Hair (like that shown above) becomes dull and coarse when the outer cuticle is ruffled (shown left).

Hair is most healthy and strong when it is slightly acidic. Using the pH scale of acidity (pH 7 is neutral, pH < 7 is acidic and pH > 7 is alkaline) hair should be about pH 4 or 5.

A solution of detergent in pure water is slightly alkaline and therefore washing hair with detergent alone will make it alkaline. This has the effect of breaking some of the crosslinks between the protein fibres in the hair. This in turn results in the outer surface of the hair, called the **cuticle**, becoming ruffled. This makes the hair look dull and coarse (figure 14).

The cuticle can also be damaged by bleaching, dyeing, curling and blow drying. Hair conditioners work by readjusting the pH of the hair to its normal value and smoothing the ruffled cuticle.

Clearly the ideal hair care involves keeping hair clean without upsetting the fine balance of the acidity of the hair and the crosslinks between the protein fibres which can result in damage to the cuticle.

Figure 13 *Changing crosslinks from wet to dry hair*

1 Write two or three sentences which explain why hair conditioners are usually slightly acidic.

2 Make a list of the different qualities which manufacturers claim for their shampoos. Use the information in this section to help you to suggest which ingredients in shampoos give them special properties.

1 Keeping clean is an important aspect of keeping healthy. Improvements in the way in which people keep themselves and their clothes clean have helped to improve public health. For example, diseases such as cholera and typhoid which are carried by water are now extremely rare in this country.

2 Cleaning your skin, your hair and your clothes involves dislodging particles of dirt or dead skin and removing oil and grease. The main problem is oil and grease which cannot be removed by water on its own.

3 For water to be an efficient cleaner it needs help. What is used to help it depends on what is being cleaned.

To clean	we use
skin	water + soap
hair	water + shampoo
clothes	water + washing powder
dishes	water + washing-up liquid
teeth	water + toothpaste

4 Soap, washing powder, washing-up liquid, shampoo and toothpaste all contain detergents. A detergent is any chemical which can be used for cleaning. Manufacturers use different chemicals as detergents in different products. For example, you would not want to use washing powder on your face because it would remove too much grease and make your skin dry and sore, but it is ideal for clothes.

5 Detergents work by:

◆ increasing the wetting power of water,
◆ breaking oil and grease down into tiny droplets which then mix with the water.

A mixture of oil and water is called an **emulsion** and because detergents help the mixture to form they are called **emulsifiers**.

6 Many cosmetics are emulsions of oil and water. The oils in cleansing cream absorb the oil and grease on our skin. When the cleansing cream is wiped off it brings the oil, grease and dirt with it.

7 There are soapy and soapless detergents. Normal tablets of soap contain the soapy type and most other cleaning substances contain soapless detergents. Soapy detergents are made from animal fats and vegetable oils. Soapless detergents are made from crude oil.

8 Tap water in some parts of the country, because it has flowed over or through particular types of rocks, contains substances dissolved in it. Some of these substances have the effect of making it difficult to form a lather with soapy detergents. The water is called **hard** water. It is dissolved calcium and magnesium compounds in the hard water which react with the soapy detergent to form a scum and so stop it forming a lather.

9 The build-up of scale in boilers, hot water pipes and kettles means that more energy is needed to heat the water. Sometimes it is necessary to spend money softening the water.

On the other hand, however, hard water is probably healthier for us because of the dissolved salts in the water. Hard water is also preferred for some industries such as brewing.

10 Detergents are essential ingredients in shampoos and toothpaste but there are also other ingredients to meet the special needs of cleaning hair and of cleaning teeth (figure 15).

Figure 15 *The ingredients needed in shampoo and toothpaste to make them effective*

Shampoos are designed to be neutral (pH = 7) or slightly acidic. They must not be alkaline (pH more than 7) as alkali makes hair break easily.

Toothpastes are usually slightly alkaline. The alkali neutralises the acids formed by the action of mouth bacteria on food.

11 Household cleaners for floors, tiles, windows and ovens usually contain grease removers, polishing agents, colouring and perfume. Oven cleaners often use strong alkali as a grease remover. Detergent action is partly responsible for the grease removal but sometimes solvents are also present which are substances which dissolve the grease.

1 Why is it difficult to form a lather with soap and hard water?

If water which comes out of your tap contains certain dissolved substances from rocks, then it will be hard water. You will know it is hard because it will be difficult to make a lather with soap. Also if your water is hard a lot of scale will appear inside your kettle.

When you add soap to hard water it is used up as it reacts with the dissolved substances in the water to form an insoluble scum. This is a chemical reaction:

$$\text{soap} \; + \; \substack{\text{substances in} \\ \text{hard water}} \; \rightarrow \; \text{scum}$$

This means that you have to keep on adding soap until all the dissolved substances are used up before any soap is left over to form a lather.

The bar chart (figure 16) shows how the amounts of dissolved substances, scum and lather present change as more soap is added to hard water.

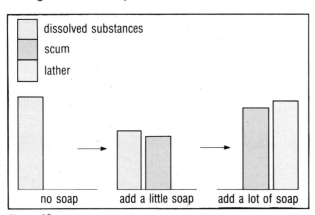

Figure 16

2 What causes hardness and where does it come from?

Observations about hard water.

◆ If a few drops of hard water are evaporated on a microscope slide, a small solid deposit is left on the slide.
◆ Water which has only flowed over rocks containing calcium or magnesium compounds is hard.
◆ Water which has only flowed over other rocks such as granite, slate or sandstone is not hard.

Observations such as those described above suggest that hardness is caused by calcium or magnesium compounds in rocks dissolving in the water.

The rock gypsum contains calcium sulphate which does dissolve slightly in pure water and so it is one cause of hardness.

Limestone-type rocks are made of calcium carbonate and sometimes a mixture of calcium and magnesium carbonate. They do not dissolve in pure water. However, these rocks are much more common than gypsum and the water in limestone areas is hard. The explanation is that limestone rocks will dissolve in rain water because the rain water contains a small amount of dissolved carbon dioxide which makes it slightly acidic (figure 17).

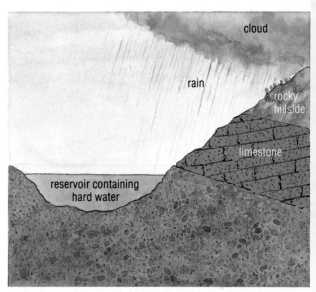

Figure 17 *Rain water falling onto a limestone area will dissolve small amounts making the water collected in the reservoir hard.*

Taking it further

When carbon dioxide dissolves in rain water a weak solution of carbonic acid is formed:

$$\substack{\text{carbon} \\ \text{dioxide}} \; + \; \text{water} \; \rightarrow \; \substack{\text{carbonic} \\ \text{acid}}$$
$$CO_2 \; + \; H_2O \; \rightarrow \; H_2CO_3$$

This solution reacts with the calcium carbonate in the limestone to form a solution of calcium hydrogencarbonate.

$$H_2CO_3 \; + \; CaCO_3 \; \rightarrow \; Ca(HCO_3)_2$$

It is the solution of this compound which makes the water hard.

The calcium part of the compound reacts with soap to form a scum:

$$\text{calcium in solution} \; + \; \text{soap} \; \rightarrow \; \text{scum}$$

3 How can hardness be removed?

To make hard water soft the dissolved calcium or magnesium has to be removed. When it has been removed the water is soft and even a small amount of soap added to it will form a lather.

When hard water is boiled an insoluble deposit is formed and the water becomes softer. After using a kettle with hard water for sometime an insoluble deposit, called **scale**, builds up inside it. This scale contains the calcium and magnesium which was previously dissolved in the water.

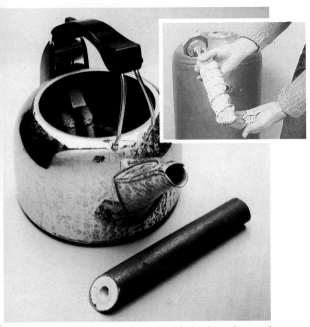

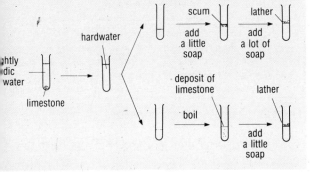

Figure 18 *Hard water scale builds up in kettles, pipes and immersion heaters and can eventually make them useless.*

Figure 19

Boiling gets the limestone out of the water (figure 19).

Another way of removing hardness from water is to add washing soda (figure 20). This is sodium carbonate. When crystals of sodium carbonate are coloured and perfumed they are sold as bath salts.

When bath salts are added to hard water they convert the dissolved calcium compound to an insoluble substance and so the water becomes soft.

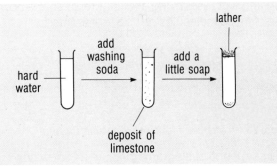

Figure 20

Taking it further

When the sodium carbonate (washing soda) is added to hard water it dissolves (figure 21).

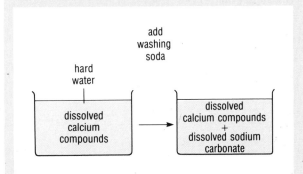

Figure 21

The carbonate part of the sodium carbonate reacts with the calcium part of the dissolved substance to form insoluble calcium carbonate – we have got the limestone back again (figure 22).

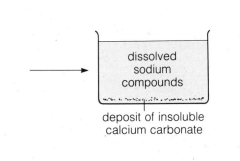

Figure 22

The sodium part still stays in the solution but as it does not react with soap it does not make the water hard.

143

4 How do detergents work?

Detergents are able to mix with both water and grease. Detergent molecules have a water-loving end and a grease-loving end. Figure 23 shows how one of these molecules can be represented.

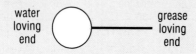

water loving end ⟵ ⟶ grease loving end

Figure 23 *A detergent molecule in diagrammatic form*

The sequence of diagrams in figure 24 show how detergent particles help remove the grease from dirty clothes to mix with water.

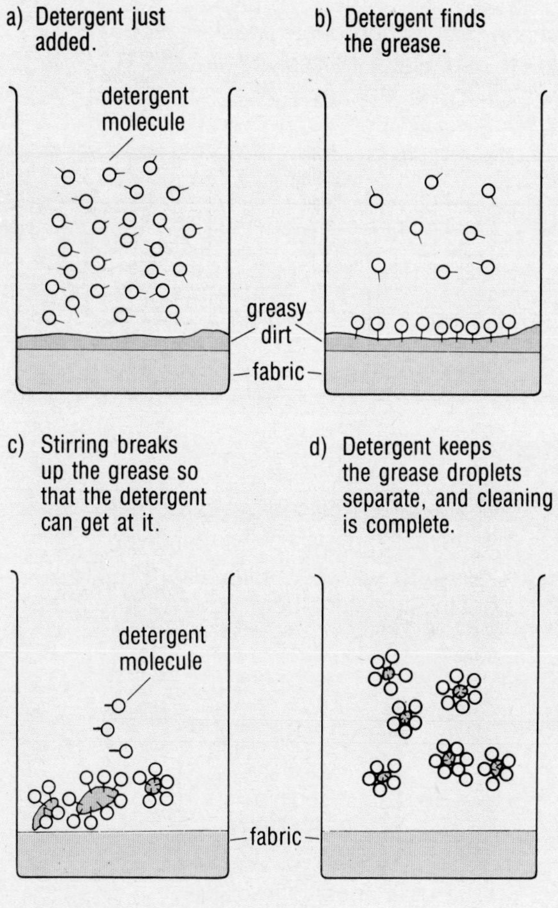

a) Detergent just added.

b) Detergent finds the grease.

detergent molecule

greasy dirt
fabric

c) Stirring breaks up the grease so that the detergent can get at it.

d) Detergent keeps the grease droplets separate, and cleaning is complete.

detergent molecule

fabric

Figure 24 *How a detergent works*

Once the grease has been broken up by the detergent, it spreads through the water as tiny droplets. This is an **emulsion** (see page 114), and it makes the water look cloudy. The grease-water emulsion is washed away when the clothes are rinsed.

Soap is the commonest detergent. But soaps are only one type of detergent. Washing-up liquid and the detergent powder used for cleaning clothes are other detergents. They are called soapless detergents. Soapless detergents are better than soap in places where the water is hard. This is because they do not form a scum with hard water.

Taking it further

As we have discussed above, detergent molecules have two parts – a water loving 'head' and a grease-loving 'tail'. What are these 'heads' and 'tails'?

Soap is made from natural oils and fats. The structures of some of the fats were described in the chapter on Food. They are formed from glycerol and a fatty acid. When fats are boiled with sodium hydroxide, they break down into glycerol and the sodium salt of the fatty acids. This salt is a soap. One example is sodium stearate (figure 25).

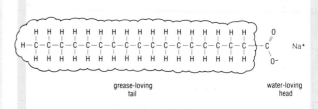

grease-loving tail — water-loving head

Figure 25 *Sodium stearate molecule*

Chemists have imitated soap, producing molecules which have similar structures but improving certain of the characteristics of soap. For example, these synthetic compounds are designed to prevent scum being formed. One detergent is shown in figure 26.

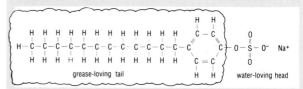

grease-loving tail — water-loving head

Figure 26 *A soapless-detergent molecule*

Unlike soap, this detergent does not react with the calcium in hard water to form a scum. This means that when a soapless detergent is added to hard water, it forms a lather immediately.

The compound which gives the grease-loving tail is produced by a series of chemical reactions from hydrocarbons. These hydrocarbons are obtained from crude oil.

5 We use emulsions everyday

If you shook up a mixture of olive oil and water in a beaker and then allowed it to stand, you would expect the mixture to separate into two layers (figure 27).

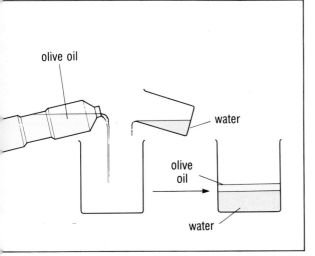

Figure 27

If you then added egg yolk to the mixture, an emulsion would be formed, known as mayonnaise (see page 115).

In the same way a detergent is also an emulsifier. The detergent in the washing-up liquid in figure 28

breaks the oil up into little droplets. The droplets then spread through the water so that it looks as though the two liquids have mixed just as the oil is broken into droplets in mayonnaise.

The mixture is an emulsion. The detergent has helped the water and oil to form an emulsion. It is therefore called an emulsifier (see page 114).

When you keep clean you depend on detergents to convert the grease and oil on your body into an emulsion of tiny droplets of the oil suspended in the water. The emulsion can then be washed off your skin or hair.

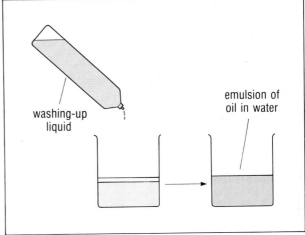

Figure 28

Things to do

Things to try out

1 If your tap water is hard there will be some scale inside your kettle. Without damaging the kettle remove a small piece of scale. Test the scale by adding it to a small volume of vinegar in a cup and observing what happens. If you can find small pieces of local stone, test them in the same way to see if they are limestone.

2 Using pH paper test the pH of any hair treatment products which you and your friends use. Record the pH numbers in a table and label each one appropriately as acidic, slightly acidic, neutral, slightly alkaline.

   Bearing in mind that hair is naturally slightly acidic, which of the products are potentially the most harmful to hair?

   In addition to shampoos try products used in bleaching, colouring, tinting, permanent waving, straightening and conditioning.

Things to find out about

3 What are the names of some of the oils which are used to make soap? Which countries do these oils come from?

Things to write about

4 Your school has decided to bury a 'time capsule'. If future generations find the capsule it should provide a picture of life in the UK during the second half of the 20th century.

   Write an account of 'Keeping Clean in the UK' for inclusion in the capsule. The account should be concise but informative. You might find it helpful to think of what we take for granted now which 100 or 200 years ago people would not have done.

Making decisions

5 A brewery company has decided that it wants to build a new brewery as the consumption of beer has gone up. A good supply of hard water is needed to make beer. Consider the possible sites marked A, B and C on the map (figure 29). Decide, from the point of view of the water supply, which of the sites is the best for the new brewery. Give reasons for your choice. Remember in real life other factors as well as the water supply would need to be considered. Decide which other factors might influence the decision.

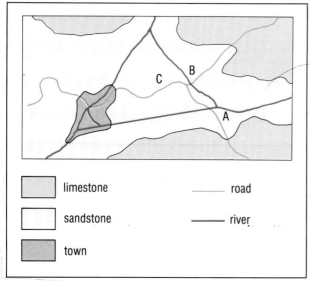

limestone   —— road

sandstone   —— river

town

Figure 29

Points to discuss

6 Some cosmetic products are used to keep our skin and hair clean and healthy. Others are used for decorative purposes. The manufacturers of the chemical substances which are used for decorative purposes use advertising to promote the sales of their products and so increase the production of the chemicals. Some of you may consider this to be a good thing others may not.

Without at first making any judgements, 'brainstorm'★ points, factors, consequences which might have some bearing on this issue. Then discuss each idea you have had in more detail.

★'*Brainstorming*' is where everyone mentions whatever comes to their minds which might have any connection with the issue. All the ideas are written down without anyone commenting on them. Then they are all looked at again to see which are the most important.

Questions to answer

Questions 7–11

A   detergent
B   emulsion
C   sebum
D   fluoride
E   scale

From the list of names, **A** to **E**, select the one which is

7 the name given to a mixture of oil and water.

8 added to toothpaste to **strengthen** teeth.

9 formed in a kettle when hard water is boiled.

10 the oily substance on hair.

11 a general name which includes soap.

12 A chemist employed by the Water Board collected samples of water from different places in her area. She tested a portion of each sample with soap solution in order to compare their hardness. She then boiled another portion of each sample, allowed them to cool and tested their hardness in the same way. The results are given in the table below.

Sample of water	A	B	C	D
Soap solution needed in cm³	0.5	5.1	3.0	1.9
Soap solution needed for boiled water in cm³	0.5	4.5	0.8	1.9

a) Which sample of water is the hardest?
b) Which sample of water is the softest?
c) Which sample(s) will appear cloudy after boiling?
d) Which sample of water will leave the most scale in kettles?
e) The local power station needs completely soft water to use in its boilers. Which of the samples would cost the most money to soften before sending to the power station?
f) Write a word equation and/or a balanced equation for the formation of scale in a kettle or water boiler.

Introducing buildings

Our homes are the most important buildings in our lives – we preserve them, insulate them and decorate them. What are our homes made from and why do we use the chosen materials?

Look at the photographs of different homes in figures 1-5 and decide in which part of the world you might find each of them. Try to identify the materials used for the roofs, the walls and the floors of these homes.

▲ Figure 1

Figure 2

◄ Figure 3

▼ Figure 4

Figure 5

In this chapter you will see that

◆ the materials for buildings are chosen because they have certain properties,
◆ the study of chemistry can help you to understand these properties,
◆ an understanding of these properties can lead to better methods of protecting buildings from the weather and to the design of new building materials.

1   Looking at building stones

Our towns and villages contain many fine buildings which have been built using a wide variety of different materials. Our study of building materials should lead you to look at these buildings with a fresh eye.

Nowadays, most buildings are constructed from bricks and concrete, which are a kind of *artificial* stone. In the past, *natural* stone was used. The most important natural building stones in Britain are sandstone and limestone, although granite and slate are also used in a few areas.

Sandstone covers a large area of Britain. The regions coloured red and yellow on the map in figure 6 are predominantly sandstone. Red sandstone is coloured red on the map and carboniferous sandstone, also known as 'millstone grit', is coloured yellow.

The name 'millstone grit' comes from the eighteenth century when corn mills used it for their grinding stones. It makes an excellent building material. It is very hard-wearing. However, its dark grey colour is rather drab and its coarse grains hold soot and dirt strongly.

Red sandstone, on the other hand, is a very pleasant colour and being softer, carves more easily. The major disadvantage with red sandstone is that it is more easily eroded as you can see from the photograph of Kenilworth Castle in figure 6.

Limestone is the most important 'traditional' building stone in Britain. It is easy to cut and carve, and its colour is varied and attractive. However, it is very easily attacked and eroded by **acid rain**.

1 What properties make millstone grit an excellent building stone?
2 What are the disadvantages of using millstone grit?
3 What advantages does limestone have as a building material compared with millstone grit?
4 What is the main disadvantage of limestone as a building material?

Glasgow tenement building

Cottages in the Lake District

The Landsdowne Crescent, Bath

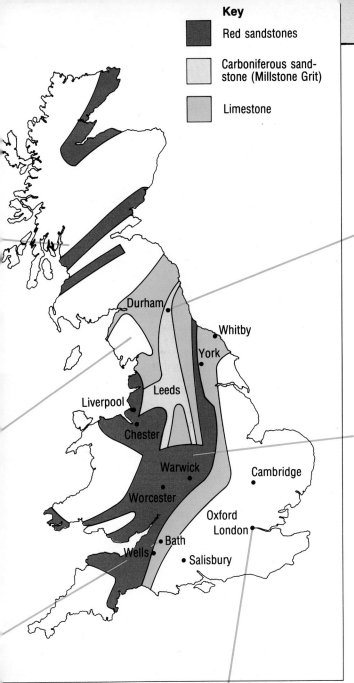

Durham
Whitby
York
Leeds
Liverpool
Chester
Warwick
Cambridge
Worcester
Oxford
London
Bath
Wells
Salisbury

Figure 6 *The areas where red sandstone, carboniferous sandstone and limestone are found in Britain.*

Durham Cathedral

Kenilworth Castle

St. Paul's Cathedral

5 St. Paul's Cathedral is built from limestone quarried near the coast at Portland in Dorset. How do you think it was brought from Dorset to London?

6 The town or city in which each of the following buildings is to be found is marked on the map in figure 6. What do you think the likely building material might be for each of them?
a) Wells Cathedral, Somerset
b) Castle Howard, Nr Whitby, Yorkshire
c) Chester Cathedral
d) Buckingham Palace
e) Liverpool Station
f) Salisbury Cathedral
g) Durham Cathedral
h) York Minster
i) King's College Chapel, Cambridge

2   How can we protect stone?

Some of the greatest works of art in the world have to withstand the weather and other atmospheric changes. Buildings and statues made of materials like limestone (such as those in figures 7 and 8) are attacked by frost and by the acids in rain. Can chemists help to prevent this damage?

Bricks and stones are made up of tiny grains with holes and pores between them. These holes and pores vary in size from invisible pores to holes which are easy to see when the material is broken. Water can be absorbed into the pores and damage occurs when the water freezes. Freezing causes the water to expand, forming ice. This will eventually crack the brick or the stone and cause bits to break off.

The acids dissolved in rain water also cause damage to buildings and this is described on pages 155–157.

Stone can be cleaned by washing it (figures 10-12). However, once the stone is clean, it must be protected from frost and acid rain. This means finding a substance which can be applied to the stone to stop the water getting into the pores – a **water repellent**.

▲ **Figure 7**   The front of the Parthenon, The Acropolis, Athens – this famous building has been seriously damaged by the weather and atmospheric pollution and is now being extensively repaired.

▲ **Figure 8**

Figure 9   What material was used to build Stonehenge? (It has been standing for over 3000 years and yet it shows very little sign of deterioration.) ▶

▼ **Figure 11** A high-speed water jet being used for cleaning a building

▲ **Figure 10**   The Liver Building, Liverpool – before cleaning

▲ **Figure 12**   The Liver Building after cleaning

What properties should a water repellent have?

- It should be insoluble in water.
- It should be a solid under all weather conditions.
- It should be easy to apply to brick or stone.
- It should not react with the brick or stone or with air.

What other properties do you think a water repellant should have?

Waxes and silicones

Chemists have looked for suitable water repellants. They have tested polymers which can be dissolved in a solvent and then sprayed on buildings. A natural polymer – **wax** – has been tried. One synthetic polymer which has also been tried is a **silicone**.

What are silicones and waxes? Silicones and waxes are large molecules. Waxes have a 'backbone' of carbon atoms to which are attached hydrogen atoms. The waxes are examples of hydrocarbons, like the one in figure 13.

Figure 15 *This jacket is water-proofed using wax. The wax stops water penetrating the fabric. Research chemists are trying to find a suitable material to achieve the same effect on stone.*

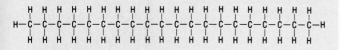

Figure 13 *A hydrocarbon molecule in a wax*

Silicones have 'backbones' made of silicon and oxygen atoms. These are the same atoms as those in asbestos and glass. One example of a silicone is drawn in figure 14. The groups of atoms attached to the silicon atoms are called **methyl** groups. They have the formula CH_3. Each methyl group has one carbon atom and three hydrogen atoms.

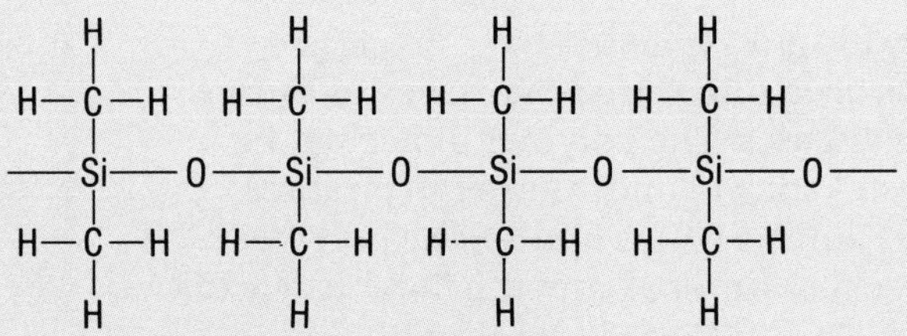

Figure 14 *Part of the structure of a silicone*

Waxes and silicones have no attraction for water and do not mix with it. So water just runs off them – they are water repellant. There are, however, several problems in coating the surface of a building:

- water is trapped in the stone behind the skin and erosion can continue,
- most waxes and silicones which have been used to protect buildings are very **viscous**. This means they cannot get into the pores in the brick or stone. One polymer, which is not so viscous, perfluoropolyether, has been used very successfully. It has been sprayed on cathedrals in Italy. However, the polymer is very expensive.

Another way of coating stone with a skin of polymer is to treat the stone with a monomer rather than a polymer. The monomer, being a smaller molecule, could get into the pores of the stone. Then the monomer molecules could be made to react with one another to form a polymer. This polymer would then fill the innermost pores of the stone as well as the outside.

So far, little progress has been made with this method of treatment. What do you think are the reasons for this?

3 Asbestos in buildings

In recent years, new Government regulations have insisted on the removal of asbestos from schools and other buildings (figure 16). Why was it necessary to remove the asbestos? Why was asbestos used in the buildings in the first place? Can we find other safer materials to use instead of asbestos? By the end of this section you will know some of the answers to these questions.

The advantages of asbestos
Asbestos is **fireproof**, **strong** (figure 17) and **unreactive**. It is mined in many countries and is widely available as a building material. There are several different minerals which we call asbestos and they are all **silicates**. Silicates are compounds with a 'backbone' of alternate silicon and oxygen atoms (figure 18).

Figure 18 The 'backbone' of a silicate

In one type of asbestos, known as **amphiboles**, the simple skeleton of silicon and oxygen atoms form long chains with bonds to metal atoms such as magnesium along the structure (figure 19). Amphiboles are examples of a **chain silicate**. Brown and blue asbestos are chain silicates.

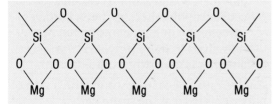

Figure 19 The structure of one form of asbestos

The silicon and magnesium have combined with as much oxygen as they can and the bonds between the atoms are very strong. This makes asbestos very unreactive and it cannot be burned. It is fireproof. This property is a great advantage when using a material in large quantities in a building. It is much safer than using something which can burn.

◀ **Figure 16** Asbestos being removed from a building. Why are the workers wearing so much protective clothing?

▲ **Figure 17** Asbestos mine at Aminadros, Cyprus

Figure 20 Asbestos lagging on pipework ▶

Another property of asbestos is that it does not conduct heat very easily. It is an insulator and has been used in buildings to insulate hot water pipes, to make roofing tiles and to make sheets for wall insulation (figure 20). Asbestos can also be shaped easily to fit around awkward structures. Not many building materials are strong, fireproof, insulators and easily shaped.

The dangers of asbestos
It has been discovered that asbestos causes lung diseases. The polymer molecules in asbestos stick to one another to produce fibres but the bonds *between* the polymer molecules are weak and the fibres can break apart. These fibres fill the air with tiny particles of dust when the asbestos is moved

a) Large	b) Medium	c) Small
asbestos fibres – too big to reach the lungs. These are trapped by the hairs in our noses.	asbestos fibres – the dangerous ones. These get into the lungs and get stuck in tiny air passages.	asbestos fibres – these are so small that they are easily cleared from the lungs.

▶ **Figure 21**
Asbestos fibres

or processed. They are small enough to get into our lungs but large enough to get stuck there. They are so unreactive that they are not attacked and destroyed by any of the substances in our bodies (figure 21).

During the 1950s and 1960s, it became clear that many people who had worked in the asbestos industry for long periods had lung diseases. These diseases made breathing so difficult that some people could not stand up, let alone walk. Even worse, some developed lung cancer and other forms of cancer. The use of asbestos was therefore strictly controlled and manufacturers of asbestos looked for safer substitutes.

But remember: *any* fine dust or smoke particles which get trapped in the tiny passages of our lungs can cause disease. This is why people who smoke sometimes get lung cancer.

A safer asbestos
Manufacturers have tried to find an alternative to blue and brown asbestos. They have looked for a substance which is strong, fireproof, an insulator and safe for us to use. They have tried to find a material which has an unreactive skeleton of silicon and oxygen atoms like asbestos. However, the most important property that the material must have is that it is safe to use.

One material which has been considered is a special form of asbestos, known as **chrysotile**. This is a silicate, but instead of the silicon and oxygen atoms being joined together in chains, as they are in blue and brown asbestos, they are formed into flat sheets.

The chrysotile sheet does a special trick. It rolls up like a carpet and forms a tube. A piece of rock taken from the ground is made up of millions of these tubes packed closely together, each tube being about 100 atoms across but millions of atoms long.

Chrysotile has all the useful properties of blue and brown asbestos but it is attacked by acids. Because of this, it is believed that it may not be such a health risk. It can be attacked and broken down by the acids in our bodies. At present, work is being done on this material by chemists and doctors.

A substitute for asbestos
A better solution may be to find building materials that do not produce dust particles so easily. The material which is often used for insulating roofs is an example. This is **glass fibre** (figure 22). It is made from a silicate and so has all the strength of the silicon-oxygen bonds found in asbestos.

These glass substitutes for asbestos are often called **Man-Made Mineral Fibres** or **MMMF** for short. The fibres are drawn out from molten glass and the size of the fibres can be controlled in the manufacturing process. If glass fibres break when they are being used they are either too large to reach the lungs (they are filtered off by our noses) or they are too small to remain trapped in our lungs.

However, there are still risks with asbestos substitutes. There are strict rules for companies which make or use asbestos and MMMFs. Anyone who is exposed to the dust must wear a face-mask (figure 23). You should also wear a mask if you are using them in the home.

Figure 22
Glass fibre production ▶

Figure 23
Anyone using MMMFs should wear protective clothing like this. ▼

1 Building materials are made from

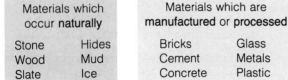

Materials which occur **naturally**	Materials which are **manufactured** or **processed**
Stone Hides	Bricks Glass
Wood Mud	Cement Metals
Slate Ice	Concrete Plastic

2 Buildings are usually made from materials that are:

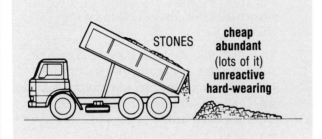

STONES

cheap abundant (lots of it) **unreactive hard-wearing**

Figure 24

3 The uses to which we put a building material depend on its properties. The properties of the material are determined by its structure. The structures of many building materials can be classified as one-, two-, or three-dimensional as shown below.

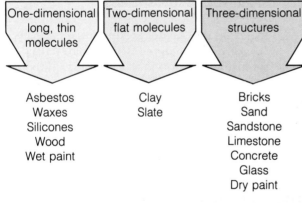

One-dimensional long, thin molecules	Two-dimensional flat molecules	Three-dimensional structures
Asbestos	Clay	Bricks
Waxes	Slate	Sand
Silicones		Sandstone
Wood		Limestone
Wet paint		Concrete
		Glass
		Dry paint

4 Bricks are made by **firing** (heating) clay to a high temperature. This causes a permanent change in the clay.

5 In many buildings, bricks are held together with **cement mortar**. This is made from limestone, clay, sand and water.(figure 25).

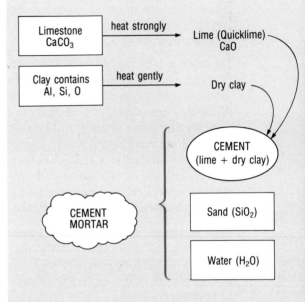

Limestone CaCO₃ — heat strongly → Lime (Quicklime) CaO

Clay contains Al, Si, O — heat gently → Dry clay

CEMENT (lime + dry clay)

CEMENT MORTAR

Sand (SiO₂)

Water (H₂O)

Figure 25 *The production of cement mortar*

6 Building materials in some areas are badly attacked by frost and by the acids in rain water.

7 The rates of chemical reactions are affected by:
   a) the surface area of the reactants,
   b) the concentration of the reactants,
   c) the temperature of the reactants,
   d) the use of catalysts.

8 Glass is made from sand, limestone and soda ash. Different types of glass can be made by adding other chemicals (figure 26).

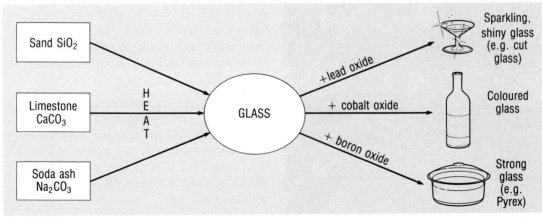

Sand SiO₂

Limestone CaCO₃

Soda ash Na₂CO₃

HEAT

GLASS

+lead oxide → Sparkling, shiny glass (e.g. cut glass)

+ cobalt oxide → Coloured glass

+ boron oxide → Strong glass (e.g. Pyrex)

◄ **Figure 26** *The production of different types of glass*

1 How are bricks made?

More bricks are used in buildings than any other material. Bricks are made from **clay**. Clay contains aluminium, silicon and oxygen atoms, linked together in separate flat layers, rather like layers of chicken wire piled up on one another. These flat layers are described as **two-dimensional structures**. When clay is wet, water molecules get between the layers and allow them to slide over one another (figure 27). This is why wet clay is soft and slippery.

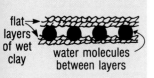
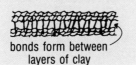

Figure 27 *The structure of wet clay*

Figure 28 *The structure of fired clay*

When clay is left to dry, most of the water between the layers evaporates. The clay also shrinks, the layers get closer and they cannot move over each other so easily. At the same time, the clay loses its slippery feel and gets harder.

When the clay is fired, it forms **brick** which is hard, gritty and rigid. During firing all the water molecules are driven out of the clay and chemical changes occur. Atoms in one layer form bonds with atoms in the layers above and below. The cross linking between the layers gives the bricks a very hard **three-dimensional structure** (as shown in figure 28). This process cannot be undone. It is not possible for bricks to be converted back to clay when water is added. Your house is safe in the rain!

Figure 29 *This furnace provides the heat necessary to produce bricks from wet clay.*

2 What is the structure of bricks and stones?

Most of the stone buildings in Britain are made of limestone or sandstone. A good way to understand the structure of sandstone – and bricks – is to start by looking at sand.

Sand is almost pure silicon dioxide (SiO_2). It has a three-dimensional structure with every silicon atom bonded to four oxygen atoms (figure 30).

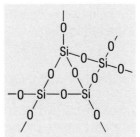

Figure 30 *The three-dimensional structure of sand* ▶

This three-dimensional structure, with strong bonds between silicon and oxygen atoms, makes sand *very hard*.

Sandstone and bricks consist of small grains of silicon dioxide and silicates with similar three-dimensional structures. In some sandstones, the grains are held together ('cemented') with silicon dioxide itself. In other sandstones, the 'cement' is calcium carbonate.

Limestone is composed mainly of calcium carbonate ($CaCO_3$) with strong bonds between the calcium and the carbonate particles.

These strong bonds in sandstone and limestone make them hardwearing and fairly unreactive – ideal for the outside walls of buildings.

3 Why are bricks and stones eroded?

Even unpolluted rain water is slightly acidic. Its pH is about 5.6. This is because as it falls, the rain water reacts with carbon dioxide in the air to form an acid.

'pure' water + carbon dioxide → 'rain' water
in air   (carbonic acid)

H_2O + CO_2 → H_2CO_3
pH 7   pH 5.6

Silicon dioxide (SiO_2) and silicates are not attacked by acids. This means that bricks and sandstones, in which the 'cement' is formed from silicon dioxide, are not affected by rain water.

Carbonates, however, are readily attacked by acids. So, limestones (which are mainly calcium carbonate) and sandstones in which the 'cement' is calcium carbonate *are* affected by the acids in rain water.

$$\text{calcium carbonate} + \text{carbonic acid in rain water} \rightarrow \text{calcium hydrogencarbonate}$$
$$CaCO_3(s) + H_2CO_3(aq) \rightarrow Ca(HCO_3)_2(aq)$$

The product is soluble in water. Limestone slowly reacts with rain water and the product is washed away.

In the case of sandstones with a 'cement' of calcium carbonate, the 'cement' slowly dissolves and the stone crumbles without anything to hold the silicon dioxide grains together.

4   How does acid rain form?

Coal and oil are the two most widely used fossil fuels. Coal is mainly carbon. Oil is a mixture of hydrocarbons (compounds of carbon and hydrogen). Both also contain small amounts of sulphur. When they burn, this sulphur reacts to form sulphur dioxide.

Figure 31   *One of the effects of the wastes from heavy industry is acid rain.*

$$\text{sulphur in fuel} + \text{oxygen in air} \rightarrow \text{sulphur dioxide}$$
$$S + O_2 \rightarrow SO_2$$

Sulphur dioxide reacts with rain water to form sulphurous acid.

$$\text{rainwater} + \text{sulphur dioxide} \rightarrow \text{sulphurous acid}$$
$$H_2O + SO_2 \rightarrow H_2SO_3$$
$$\text{pH 5.6} \qquad\qquad\qquad \text{pH 4}$$

Some of the sulphurous acid is oxidized to sulphuric acid (H_2SO_4). When sulphurous acid and sulphuric acid are present in rain water, the rain water has a pH between 4 and 4.5. This is much more acidic than unpolluted rain water. It is this polluted acid rain which attacks some buildings much faster than unpolluted rain water.

Taking it further

Are there other acids in our rain? The answer is yes, as we have seen already (page 155), carbonic acid is formed when carbon dioxide in the air reacts with unpolluted rain water. Sulphurous and sulphuric acids are also both found in polluted rain water.

Nitric acid is also present in rain water and is produced from oxides of nitrogen (nitrogen monoxide (NO) and nitrogen dioxide (NO_2)).

Some nitrogen oxides are formed naturally. However, these oxides are also produced in the engines of cars and lorries and are emitted in their exhaust fumes. Furnaces used to generate electricity from coal and oil also emit oxides of nitrogen. These oxides of nitrogen, together with ozone in the atmosphere, speed up the conversion of sulphur dioxide to sulphur trioxide, which then dissolves in water to produce sulphuric acid.

As you can see it is important to reduce the emission of both sulphur dioxide *and* oxides of nitrogen. It is not enough to cut down the production of one without reducing the other, if we wish to reduce the acids in rain.

5   How does acid rain affect buildings?

Acid rain has particularly bad effects on the materials commonly used in buildings: metals, limestone and sandstone.

Some metals, including iron, react with sulphuric acid producing hydrogen.

$$\text{iron} + \text{sulphuric acid} \rightarrow \text{iron sulphate} + \text{hydrogen}$$
$$Fe + H_2SO_4 \rightarrow FeSO_4 + H_2$$

This reaction causes the metal to wear away. Because of this, iron and steel are normally painted to protect them from rusting and acid rain.

The problems caused by rusting and the acids in rain water have led to the development of new building materials. For example, iron from which gutters and drainpipes used to be made is now being replaced by the plastic, PVC. This does not rust and is not attacked by acids in rain water.

At one time, steel and wood were the only materials used for window frames. Nowadays, aluminium is often used. It does not rust like iron or rot like wood. However, it is attacked by the acids in rain water.

By investigating the reactions of metals with water and with acids, it is possible to draw up a reactivity series for metals. This reactivity series is similar to that described on page 9 when the reactions of metals with water were discussed.

Sodium	**Most** reactive
Calcium	
Magnesium	
ALUMINIUM	
ZINC	
IRON	
COPPER	**Least** reactive

The metals used as building materials are written in capital letters. These are at the bottom of the list.

Sulphuric acid, in acid rain, causes other problems. As we have seen already limestone and sandstone react with rain water. Acid rain, because it is more acidic reacts faster than unpolluted rain. A further problem is that the sulphuric acid in acid rain water reacts with calcium carbonate to form calcium sulphate.

$$\begin{array}{ccccccc} \text{sulphuric} & + & \text{calcium} & \rightarrow & \text{calcium} & + & \text{carbonic} \\ \text{acid} & & \text{carbonate} & & \text{sulphate} & & \text{acid} \end{array}$$

$$H_2SO_4(aq) + CaCO_3(s) \rightarrow CaSO_4(s) + H_2CO_3(aq)$$

The calcium sulphate is insoluble in water so it forms a solid in the cracks and pores of the stone. As the solid calcium sulphate forms, it expands making the cracks even wider so that bits of stone eventually fall off and the stone crumbles away.

Figure 32 *Acid rain also has a devastating effect on plant life.*

We can investigate the effect of increasing the concentration of acid in rain water in the laboratory.

The apparatus in figure 33 is designed to find out what happens when limestone chips are reacted with different concentrations of hydrochloric acid. Look carefully at figure 34. This shows the volume of carbon dioxide produced when limestone chips are reacted first with 2.0M hydrochloric acid and then with 1.0M hydrochloric acid.

The same mass of limestone was used in both experiments and all the limestone chips were about the same size.

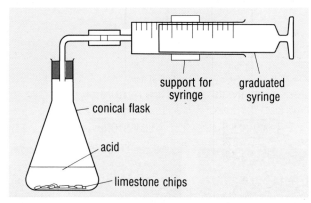

Figure 33 *Apparatus used to collect and measure the volume of carbon dioxide produced when acid reacts with limestone.*

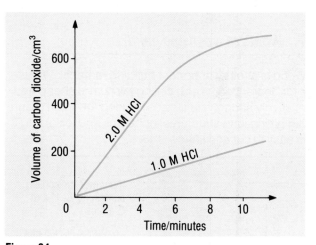

Figure 34

The results in figure 34 show that 2.0M hydrochloric acid produces carbon dioxide faster than 1.0M acid. The reaction goes faster with more concentrated acid. In the more concentrated solutions of the acid, there are more particles of acid in a certain volume. This means that there is more chance of a collision between a limestone particle and an acid particle and the reaction rate increases. Experiments using smaller pieces of limestone speed up the reaction still further.

6 What stone is a building made from?

The action of acid on buildings is often very harmful. However, we can use the effect of acids on stone in the key which follows (figure 35) to help us find out what material has been used in a particular building.

Figure 35 *Stone-identification chart*

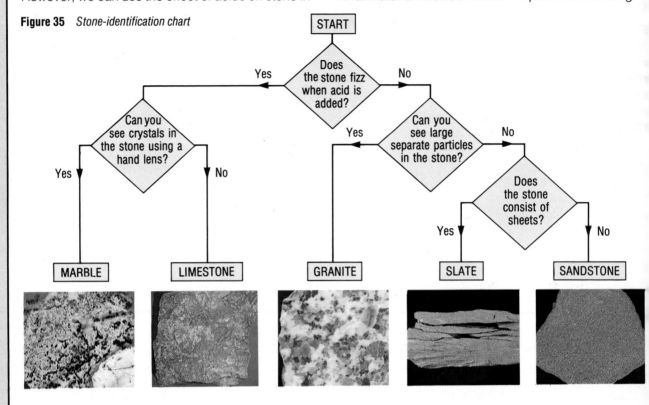

7 What are wood and paint?

Wood is another important building material. We use it for doors, beams and ceiling joists. It is cheap and relatively strong. It splits easily along the grain (figure 36) giving long structures for beams and joists and it is easy to cut and shape.

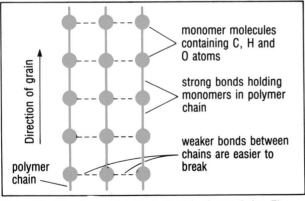

Figure 36 *Wood consists of long parallel polymer chains. The weak bonds between the chains are easy to break, so wood splits 'along the grain' – down the line of the polymer chains.*

Unprotected wood is attacked by the weather (figure 37) and insects (figure 38) and this causes it to rot.

◀ **Figure 37**
Timber being destroyed by dry rot. Is it really dry?

▲ **Figure 38**
Timber damaged by woodworm

◀ **Figure 39**
Protecting the surface from attack will prolong its life.

What is paint for?

Paint is used on wood, stone and metal to protect them. Many gloss paints contain a polymer, a solvent (which dissolves this polymer) and pigment to colour the paint.

Paint polymers contain long chains of carbon, hydrogen and oxygen atoms. Some are called **alkyd resins**. These long molecules get tangled up, which makes the paint thick. When the paint is brushed onto a surface, the molecules partly untangle. This allows them to move more freely and the paint flows more easily (figure 40).

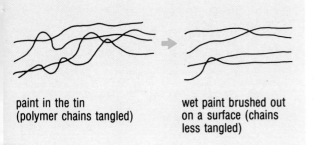

paint in the tin
(polymer chains tangled)

wet paint brushed out on a surface (chains less tangled)

Figure 40

As the paint dries, the first thing that happens is that the solvent evaporates. At the same time, the polymer starts to react with oxygen in the air. When alkyd resins react with oxygen the polymer chains get joined.

The cross links between the chains hold them together and stop them moving (figure 41). The paint hardens and sets. This hardening happens on the surface of the paint at first. Later, as oxygen reaches the paint underneath this sets hard as well.

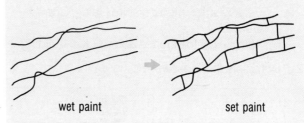

wet paint

set paint

Figure 41

The hard paint covers the surface of the wood and seals it off. In this way it protects the material from the weather and insects.

Taking it further

When alkyd resins in paint come in contact with air they begin to set. The setting process involves oxidation – a reaction between the alkyd resin and oxygen in the air.

Alkyd resins are specially-designed molecules. They have two parts:

◆ a long polymer chain containing carbon, hydrogen and oxygen atoms and

◆ side-chains containing carbon and hydrogen atoms at frequent intervals along the polymer chain. The side chains are also unsaturated – they contain double bonds between some of the carbon atoms (figure 42).

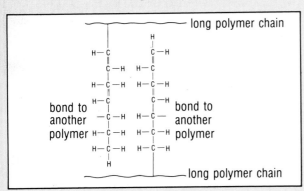

Figure 42 *The structure of alkyd resins before setting*

On exposure to the air, the double bonds in the side-chains react with oxygen. Cross links are formed between unsaturated sites on neighbouring polymers (figure 43).

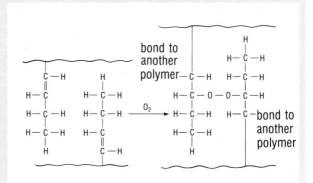

Figure 43

The cross links between polymer chains hold the paint molecules together. This prevents movement of the molecules and the paint sets and becomes hard. Figure 43 shows cross links between two oxygen atoms. However, these links may be between a carbon atom and an oxygen atom or between two carbon atoms.

Things to try out

1 Make a collection of photographs showing buildings made from different types of bricks and stones. (Magazines, advertisements, tourist leaflets and postcards are useful sources.) If possible, build up your collection by including:
   a) pictures of the scenery associated with the area from which the stone or clay came, e.g. limestone scenery on the coast (Beachy Head, Lulworth Cove, Dover) and inland (the Peak District, the Yorkshire Dales, Cheddar Gorge).
   b) samples of the stones and brick in buildings.

2 Complete the crossword in figure 44.

Figure 44

Across
1 Holds bricks together (6)
2 Reverse pat to empty the furnace (3)
7 Burns easily when cut into small pieces (3)
8 Silicon dioxide occurs naturally as this (4)
11 Mixed with limestone to make cement (4)
12 An important constituent of 'Pyrex' glassware (5)
13 Used to protect wood and metal (5)
14 and 15 (down) Glass is made from this (4, 3)
18 The symbol of one of the elements in clay (2)
19 Reacts with paint on setting (6)

Down
1 Often measured in reaction rate experiments (4)

Down (cont.)
2 Trade name of material used to protect wood from insects (8)
3 Symbol of element replacing wood in window frames (2)
4 This can be 'wet' or 'dry' in wood (3)
6 Alkyd resins in paint are this kind of substance (7)
9 The kind of lime in lime mortar (6)
10 Pollutant which dirties buildings (4)
14 Prehistoric tribes hardened clay in this to make pots (3)
15 See 14 across (3)
16 Symbol of an element whose oxide is used to colour glass (2)
17 Symbol of a poisonous element used in thermometers (2)

3 The names of ten building materials are hidden in the wordsearch below. How many can you find?

F	A	Y	J	E	M	I	O	D	G	E	D
I	B	P	O	S	T	E	E	L	N	R	B
K	P	E	D	C	O	H	T	U	L	A	I
C	A	L	M	I	A	L	E	N	O	T	S
H	G	I	A	W	E	B	R	A	E	K	C
G	L	A	S	S	D	O	C	F	O	Y	G
N	I	M	U	F	T	E	N	W	A	R	M
A	P	O	C	I	J	I	O	L	E	T	U
E	N	H	M	B	A	M	C	C	O	G	D
L	U	B	G	E	V	A	N	S	H	D	A
F	E	O	K	C	I	P	L	V	B	J	L
R	C	R	E	D	W	T	H	A	T	C	H

Things to find out

4 The porous nature of bricks means that water can be absorbed into a building from the ground or from driving rain.
   a) How do builders prevent absorption of water into a building from the ground?
   b) Why are the problems from driving rain greater in old buildings with thick walls than in modern buildings with cavity walls?
   c) What problems result when the brickwork of a house is permanently damp?
   d) The water absorption of bricks is expressed as a percentage increase in mass when the bricks are placed in water, i.e.

$$\text{water absorption} = \frac{\text{increase in mass due to water}}{\text{mass of dry brick}} \times 100$$

   Describe how you would measure the water absorption of a sample of brick.

5 Chalk, marble and limestone are all forms of calcium carbonate which are found naturally.
   a) How have these deposits of calcium carbonate been formed?
   b) Why are they so different in hardness?
   c) Marble is the hardest of the three but limestone is used more often for buildings. Why is this?

6 a) What is the commonest type of rock in your area?
   b) Are there any quarries (open or disused) in your area? If so, what do (did) they produce?
   c) What type of stone is used in your local buildings?

Points to discuss

7 Imagine that you belong to a local society concerned with the conservation of historic buildings in a rural limestone area. Plans have just been published showing the proposal to build a large chemical factory and a coal-fired power station in your area. A public enquiry will be held to hear objections to the scheme. Discuss what your society should say at the public enquiry.

Questions to answer

8 This question is about the corrosion of metals and stonework and the pH of rain water.
   a) What acid is present in unpolluted rain water?
   b) Explain, with an equation, how this acid gets into rain water?
   c) Explain how sulphurous acid gets into polluted rain water.
   d) Why does polluted rain water cause more damage to metals and stonework than unpolluted rain water?
   e) Name one kind of stone which is affected by rain water.
   f) What reactions take place between the stone in part (e) and the acids in rain water?

9 This question is about clay and bricks.
   a) Why is clay slippery and smooth when it is wet?
   b) What happens to the structure of clay when it is changed into brick?
   c) Why are bricks hard and rigid?
   d) What happens to bricks when they are damaged by frost?
   e) Explain how frost damages bricks.
   f) How would you compare the effect of frost damage on two different types of brick?

10 This question is about gases which pollute the air and the effect of these gases on limestone.
   a) Name three of the main gases which pollute normal air.
   b) Explain how these three gases come to be present in the air.

   c) Explain how polluted air can damage the limestone used in buildings.
   d) Outline two methods that have been used to protect stonework from decay.

11 Figure 45 shows the rate of reaction between some limestone chips and 1.0M hydrochloric acid at 25°C.

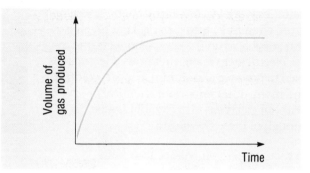

Figure 45

Redraw the graph and sketch in the lines you would expect if:
   a) the same mass of limestone powder was used. Label this line A.
   b) the temperature was 15°C. Label this line B.
   c) half the mass of limestone chips was used. Label this line C.

12 Some information on the corrosion of steel is given in the table below.

Site	Type of climate and environment	Relative rate of corrosion
Basrah, Iraq	Dry, sub-tropical, urban	1
Apapa, Nigeria	Tropical, coastal, rural	3
Llanwrytydd, Mid-Wales	Temperate, rural	7
Motherwell, Scotland	Temperate, industrial	13

   a) Why is the rate of corrosion so slow in Basrah?
   b) Why is corrosion faster in Apapa than in Basrah?
   c) Why is corrosion faster in Motherwell than in Llanwrytydd?

13 A new stone is being quarried for use as a building material. Describe the experiments that you would carry out to check the quality and suitability of the stone as a building material.

Introducing making and using electricity

These electrical appliances will only work if they are supplied with electricity. The appliances in figure 1 use the mains supply.

The appliances in figure 2 need one or more batteries.

Batteries work by converting chemical energy (from chemicals) into electrical energy. It is also possible to do the reverse process and to use electricity to make chemicals. Electricity is used to produce aluminium, copper, chlorine and many other important chemicals. Without electricity there would not be enough of these chemicals.

In this chapter you will see how

◆ chemistry and electricity are linked,
◆ electricity is produced from chemicals,
◆ chemicals are produced using electricity.

◀ **Figure 1**
Electrical appliances which operate using mains electricity.

Figure 2
Electrical appliances which use batteries. ▼

▶ **Figure 3** *What are the advantages of using batteries instead of mains electricity?*

▼ **Figure 4** *Electrolysis can be used to coat metal objects - like the copper lids and cylinders in the photograph. Why is this a useful process?*

1 Why some cells leak

Have you ever opened your radio or torch to put in new batteries and found a messy old battery inside, as in figure 5?

Many of the batteries in general use are **dry cells**, usually zinc-carbon cells. Zinc forms the outer casing and is the **negative electrode**. The **positive electrode** is made of carbon.

The battery also contains an **electrolyte** – ammonium chloride – which allows the electricity to flow between the electrodes. It is acidic and reacts with the zinc causing severe corrosion. Eventually, holes will appear in the zinc and the electrolyte may then begin to attack the appliance itself.

You can buy **leak-proofed** batteries. In these, the zinc casing is sealed in either plastic or steel.

You can also buy **non-leaking alkaline cells** (figure 6). In these, zinc is again used as the negative electrode but it is not used as the casing. Instead the zinc is inside the cell and the outer case is steel. The electrolyte is potassium hydroxide, an alkali. The positive electrode is manganese dioxide.

As with the zinc-carbon cells, zinc loses electrons to form zinc ions:

$$Zn \rightarrow Zn^{2+} + 2e^-$$

The manganese in manganese(IV) oxide accepts electrons and the reaction can be written as:

$$Mn^{4+} + e^- \rightarrow Mn^{3+}$$

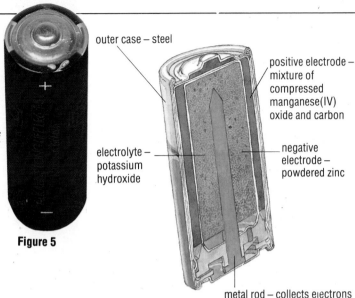

Figure 5

positive electrode – mixture of compressed manganese(IV) oxide and carbon

outer case – steel

electrolyte – potassium hydroxide

negative electrode – powdered zinc

metal rod – collects electrons

Figure 6   *Non-leaking alkaline cell*

Figure 7
The components of a battery

1 Why doesn't an alkaline cell leak?
2 Describe how you would try to test the claim made in the advertisement in figure 8.

Figure 8

2 Little cells to the rescue…!

Do you ever think that your heart might stop while you are asleep? Fortunately for most people this is not likely to happen but it is a reality for some people. They are alive only because their hearts are kept beating at a steady rhythm by small **heart pacemakers** placed under the skin near their hearts (figure 9).

Modern pacemakers contain the metal **lithium**. What an element to choose! It burns very easily in air and reacts rapidly with water to form hydrogen – a flammable gas.

However, lithium is a very reactive metal which readily loses electrons. This is a property that is very important for cells. To complete the cell another material is needed that easily accepts electrons. One that has been chosen for heart pacemaker cells is silver chromate (Ag_2CrO_4). It is the chromium atom that accepts the electrons.

The cell also needs an electrolyte which will allow the electrons to flow from the lithium to the silver chromate. The electrolyte must *not* contain any water. This problem was solved by research chemists, who decided to use a compound (lithium chlorate(VII), $LiClO_4$), which has ionic bonding, dissolved

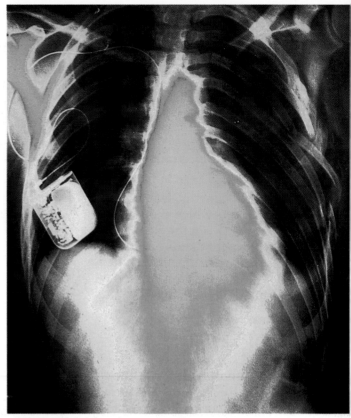

▲ **Figure 9**   *A pacemaker can be a life saver*

in the solvent propene carbonate. Using this solvent instead of water makes the overall cell safe (figure 10) even though the individual components are so reactive.

Pacemaker cells can last 10 years. People who have to wear pacemakers can sleep soundly at night, relying on lithium.

1 How do pacemaker cell designers use the electrochemical series when thinking about materials to use in cells?
2 Why couldn't water be used in the cell?
3 Pacemaker cells could be designed to be much smaller than the present button cells. What would be the advantages and disadvantages of tiny cells?

3 Power for the shuttle

The United States Space Agency, NASA, has used the same space shuttle *Discovery* (figure 11) to place satellites in space. *Discovery* is lifted off using **chemical energy**. When in orbit it releases the satellite.

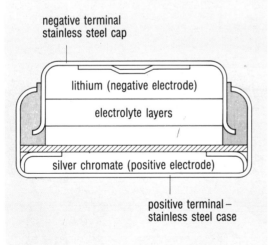

negative terminal
stainless steel cap

lithium (negative electrode)

electrolyte layers

silver chromate (positive electrode)

positive terminal –
stainless steel case

Figure 10

Figure 11 *The American space shuttle orbiter Discovery being launched in September 1988 for a successful 5 day flight.*

This spacecraft is powered by a fuel cell (figure 12). The fuel cell converts the energy released when hydrogen and oxygen react.

$$2H_2 + O_2 \rightarrow 2H_2O$$

When hydrogen burns in oxygen an enormous amount of heat is produced. It is a highly **exothermic** reaction.

Figure 12 *Electrical power for this spacecraft is supplied by fuel cells where chemical reactions are converted into electricity. Solar panels along the length of the body of the orbiter provide extra power in flight.*

In the spacecraft, instead of allowing hydrogen and oxygen to burn, an electric cell is set up. Figure 13 shows a simple version of the cell.

At the negative electrode:

$$2H_2 \rightarrow 4H^+ + 4e^-$$

At the positive electrode:

$$O_2 + 2H_2O + 4e^- \rightarrow 4OH^-$$

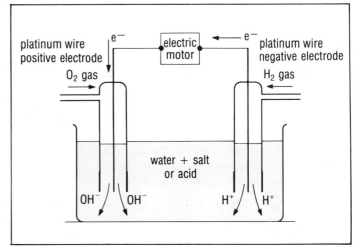

Figure 13 *A fuel cell*

Figure 14 *Space scientists body functions are continuously monitored using sensors applied to the skin.*

These fuel cells have also been the major energy cells in the moon space missions. The water produced in the cells is drunk by the astronauts.

Many research laboratories are trying to develop other fuel cells which would serve as non-polluting energy sources.

In theory, any chemical which can burn in oxygen can also work in a fuel cell. Hydrocarbons could be used if efficient catalysts can be developed. As the fuel cell would work at lower temperatures than an internal combustion engine, no nitrogen oxides would be formed. The only products would be carbon dioxide and water.

1 What other ways are there for the satellite, when in orbit, to get electrical energy?
2 Why aren't nitrogen oxides formed in a hydrocarbon-oxygen fuel cell although they are found when a hydrocarbon burns in air as in a car engine?

4   Shining surfaces from electrolysis

Are the handlebars of your bicycle bright and shiny (or were they, when the bike was new)? Handlebars are usually made of iron, which is not a particularly shiny metal. However, their *surface* is not iron but a thin layer of chromium, a bright silvery metal.

You will also see chromium on car bumpers and kettles and in many places where a gleaming metal surface is wanted (figure 15).

The only way that this thin shiny layer of chromium can be put onto duller, less expensive metals, such as iron, is by **electrolysis**. The process of coating one metal with another, using electricity, is called **electroplating**. The metals used for electroplating are those low in the electrochemical series.

The metal object to be electroplated is made the negative electrode. A piece of the metal used for plating is made the positive electrode. The electrolyte contains ions that are formed from the positive electrode.

In the example in figure 16, the positive electrode is nickel and the electrolyte solution is nickel sulphate. The positive electrode dissolves into the solution,

$$Ni \rightarrow Ni^{2+} + 2e^-$$

The nickel ions are deposited on to the copper negative electrode and plates it.

$$Ni^{2+} + 2e^- \rightarrow Ni$$

Nickel, chromium, copper, silver and gold are all used to electroplate other metals (figure 17).

Figure 15   *Chromium plated tubes and knobs*

⊕ terminal

positive electrode – nickel plate

⊖ terminal

positive electrode – nickel plate

nickel(II) sulphate solution

negative electrode – copper foil to be plated

▲ **Figure 16**

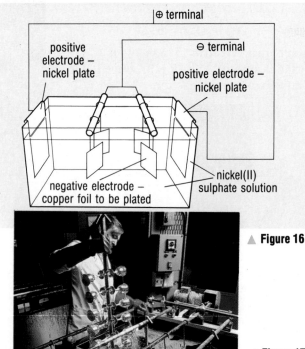

◀ **Figure 17**
Electroplating silver cups

Some metals do not stick together successfully. For example, it is difficult to electroplate chromium directly on to iron. Handlebars or car bumpers are first electroplated with nickel and after that with chromium.

A great advantage of electroplating over spray-painting is that metal objects of any shape can be completely covered. Anywhere that ions in the electrolyte solution can flow will receive a coat of the new metal. This is particularly effective for difficult areas such as the insides of hollow iron containers which must be protected from corrosive liquids.

It is not necessary for the whole object to be made of metal. Only the surface must be a good conductor of electricity. It is possible to coat objects such as leaves with conducting metallic or carbon paint, then electroplate them with copper or silver. The technique is often used in making jewellery or other decorations.

1 Your bicycle is more expensive because the handlebars are chromium-plated. Is the expense worth it? What other surfaces may be just as suitable and decorative for bicycles and car bumpers?
2 How would you make your own copper-covered ornaments, starting with objects made from wood, clay or plastic?

In brief
Making and using electricity

1 There are two types of electric charge: positive (+) and negative (−).
Opposite charges attract (figure 18a), similar charges repel (figure 18b).

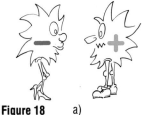

Figure 18   a)   b)

2 Conductors are materials in which electric charge can move easily when a potential difference (electric push) is applied.

3 Atoms have an internal structure. They are made of different sorts of particles (figure 19).

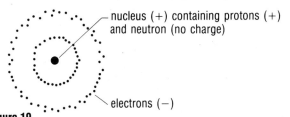

nucleus (+) containing protons (+) and neutron (no charge)

electrons (−)

Figure 19

4 Atoms can lose or gain electrons to form charged ions:

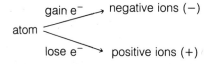

atom — gain e^- → negative ions (−)
atom — lose e^- → positive ions (+)

5 Discharge of ions involves loss or gain of electrons.

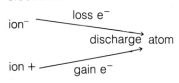

ion^- — loss e^- → discharge → atom
$ion +$ — gain e^-

6 **In elements**, electric current is a flow of *electrons* and no chemical change occurs.

Electric current is a flow of electric charge.

In compounds, electric current is a flow of *ions* and chemical decomposition occurs.

7 Compounds are mainly either ionic or molecular:

ionic: metal/non-metal compounds made up of ions held together by the attraction of opposite charges (**ionic** bonds).

compounds

molecular: compounds made up of molecules consisting of atoms joined by sharing electrons (**covalent** bonds).

8 Chemical reactions often give out energy. In electric cells the chemical energy of reactions is converted into both heat and electrical energy.

9 Electric cells contain two electrodes, one is positively charged (+) and the other is negatively charged (−) (figure 20). The electrodes are made of metal or carbon. The negative electrode releases electrons. These electrons flow through the external circuit to the positive electrode. The electrodes are placed in a solution called the electrolyte. In **wet cells** the electrolyte is an acid, alkali or salt solution in water. In **dry cells** the electrolyte is held in a paste.

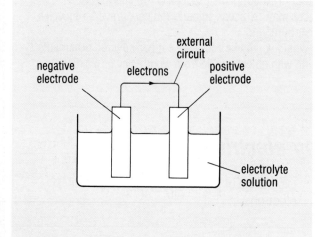

Figure 20

10 A number of cells can be joined together to make a battery. **Primary** cells cannot be recharged after they run down. A **secondary** cell, such as a lead/acid cell, can be recharged by converting electrical energy back into chemical energy.

11 The electrochemical series of metals is similar to the reactivity series. The order in the electrochemical series compares how easily the metal atoms (*M*) lose electrons to form positive ions in solution, for example:

$$M(s) \rightarrow M^+(aq) + e^-$$

K Na Ca Mg Al Zn Fe Pb Cu Hg Ag Au
electrons *most*   electrons *least*
  easily lost   easily lost

12 Some metals can form more than one type of ion. For example, copper can form Cu^+ or Cu^{2+} ions. Compounds containing these ions are known as copper(I) and copper(II) compounds.

13 The main use of electricity in chemistry is the production of chemicals by electrolysis. Electrolysis is the decomposition of a compound when electricity passes through it. The compound must be molten or dissolved in water as this allows ions to move. Compounds which conduct electricity is this way and decompose are called electrolytes.

14 During electrolysis changes take place at the electrodes. At the negative electrode: positive ions in the electrolyte gain electrons. Metals or hydrogen are formed.

$$M^{2+}(aq) + 2e^- \rightarrow M(s)$$

$$2H^+(aq) + 2e^- \rightarrow H_2(g)$$

At the positive electrode: negative ions in the electrolyte lose electrons. Non-metals are formed.

$$2X^-(aq) \rightarrow X_2(g) + 2e^-$$

15 The main uses of electrolysis are shown in figure 21.

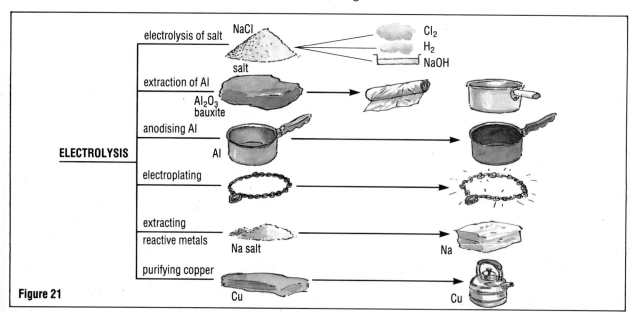

Figure 21

1 Charges and currents

If you rub a balloon with wool it will attract small pieces of paper or stick to the wall or make your hair stand on end (figure 22). It has been **charged** with electricity. Some other materials such as plastics or glass can also be charged by rubbing.

Figure 22
Static electricity drives clean hair wild! ▶

You will find that two charged balloons repel each other but attract some other charged materials. This shows that there are two types of electric charge. These have been given the names **positive** and **negative** charge.

The most important fact to remember about electric charge is that:

> opposite charges attract;
> similar charges repel.

Electric charge can move easily in materials such as metals or carbon (graphite) or solutions of acids, alkalis or salts. Such materials are good **conductors** of electricity. Charge will not flow in materials such as rubber, plastic or glass, which are **insulators**.

The flow of electric charge in a conductor is an electric current. There will only be an electric current in a conductor when there is some form of 'electric push', known as a **potential difference**, across the conductor. The potential difference is measured in **volts** and is often called 'the voltage'. Chemical reactions may be used to produce potential differences and these reactions are the basis of electric cells.

For hundreds of years the nature of electric charge and electric current was a mystery. A full explanation only came after a flurry of exciting discoveries about the structure of atoms, earlier this century.

Taking it further

Electric charge is measured in units called **coulombs**.

Electric current is measured in **amperes** – often called **amps**.

 1 amp is a current of 1 coulomb flowing per second.

Potential difference (voltage) is measured in **volts**. Energy is transferred when charge moves in a conductor and 1 volt is the potential difference when 1 joule of energy is transferred for each coulomb flowing in the conductor.

2 What is an ion?

An atom contains charged particles (figure 23). When the number of electrons (negatively charged particles) is the same as the number of protons (positively charged particles), their charges balance and the atom is electrically **neutral**.

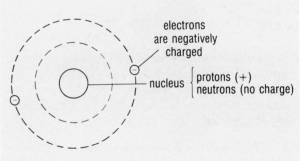

electrons are negatively charged

nucleus { protons (+)
          neutrons (no charge)

Figure 23

However, it is possible to *remove* one or more electrons from an atom, or *add* one or more electrons to it. Then the atom becomes charged.

When an atom *loses* an electron it becomes **positively** charged. The number of protons (+) is now greater than the number of electrons (−).

When an atom *gains* an electron it becomes **negatively** charged. The number of electrons (−) is now greater than the number of protons (+).

Any atom or group of atoms with a positive or negative charge is known as an **ion**. For example: when a sodium atom (Na) loses an electron, it becomes a positively charged sodium ion, shown as Na^+. When a chlorine atom (Cl) gains an electron, it becomes a negatively charged chlorine ion, shown as Cl^-.

The way in which atoms form positive or negative ions helps to explain why balloons and wool stick to each other when charged. When you rub a balloon with wool, electrons rub off the wool onto the balloon. So the balloon becomes negatively charged. The wool is now positively charged as it has fewer electrons. Remember: opposites attract – the wool and balloon stick together.

3   How do we get electricity from chemicals?

Do you have any fillings in your teeth? Have you ever bitten a piece of metal foil and experienced a sudden pain? Well you have just had a small electric shock! All that is needed to produce electric current is two different metals and a conducting solution to dip them in. Your fillings contain metal, the foil is another metal and your saliva is a conducting solution. Put these together and the nerves in your mouth let you know you have had a small shock.

Figure 24   *Ooh! Ow!*

The combination of two different metals and a conducting solution is called an **electric cell** and will produce electricity. You can make a simple cell from any pair of different metals but the liquids must be chosen carefully (table 1).

Table 1

Some liquids which are *effective* in cells	Some liquids which are *not effective* in cells
dilute acids	distilled water
dilute alkalis	ethanol
solution of salts	oil

The difference between the sets of liquids in table 1 is the presence or absence of ions. Acids, alkalis and salts all dissolve in water to give ions. Their solutions are good conductors of electricity. These compounds, which give ions when dissolved in water, are called **electrolytes**.

Distilled water, ethanol and oil contain very few or no ions. They do not allow an electric current to flow through them. They are **non-electrolytes**.

If you have tried to make simple cells from different pairs of metals, you will probably have used apparatus similar to that in figure 25.

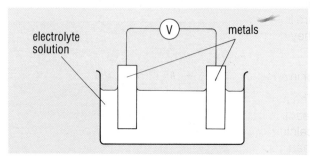

Figure 25   *A simple cell*

You will have needed to connect the metals to a voltmeter which will measure the 'electric push' – 'the **voltage**' – produced by the cell. You will have observed that:

◆ different pairs of metals give different voltages,
◆ chemical reactions appear to take place on the metal surfaces, for example, bubbles of gas appear.

4   The electrochemical series

A pattern may be found among the measurements of cell voltages for different pairs of metals. It is possible to put the metals in a series, so that the further the metals are apart in the series, the greater the voltage of the cell. The closer the metals are in the series, the smaller the voltage. This series can include all metals but, for a few well-known metals, it is as shown in the list below.

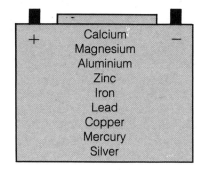

The largest voltage for a cell containing a pair of metals from this series is about 3 V from calcium (at the top of the list) and silver (at the bottom of the list).

This series, produced by the behaviour of metals in electric cells, is called the **electrochemical series**. It has a strong resemblance to another series which you will have come across when comparing the reactivity of metals – the reactivity series.

The similarity between the electrochemical series and the reactivity series is not accidental. Both series are linked by the fact that metals lose electrons when they react and form positive ions:

$$M \rightarrow M^+ + e^-$$

or in solution $M(s) \rightarrow M^+(aq) + e^-$ M = metal

The more easily a metal loses electrons the more reactive it is. This is why metals such as sodium, calcium and magnesium are much more reactive than copper or silver, i.e.

$$Mg(s) \rightarrow Mg^{2+}(aq) + 2e^-$$

occurs much more readily than

$$Cu(s) \rightarrow Cu^{2+}(aq) + 2e^-$$

A metal conducts electricity because its structure allows electrons to move about inside it.

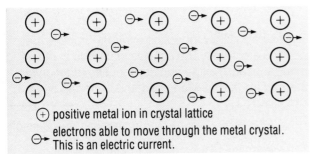

⊕ positive metal ion in crystal lattice

⊖► electrons able to move through the metal crystal. This is an electric current.

Figure 26   *The structure of a metal*

The structure of a metal consists of a lattice of positive ions surrounded by freely moving electrons (figure 26).

A current flows in a metal wire when electrons flow along the metal.

5   *What happens inside an electric cell?*

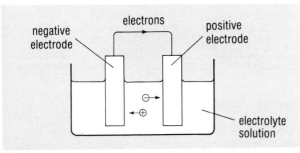

Figure 27   *An electric cell*

All electric cells in a circuit have:

a **negative electrode** which releases electrons,
a **positive electrode** which accepts electrons,
a solution which allows ions to flow through it (an **electrolyte**).

This flow of electrons and ions is the **electric current**.

A simple cell

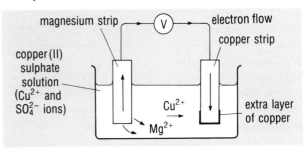

Figure 28   *A simple cell*

In the simple cell in figure 28, pieces of magnesium and copper are dipped in copper(II) sulphate solution. The metals are connected through a voltmeter and a voltage is registered. The magnesium atoms in the strip tend to lose electrons and become positive ions (Mg^{2+}). These ions dissolve into the electrolyte leaving the extra electrons behind.

$$Mg \rightarrow Mg^{2+} + 2e^-$$

These electrons move along the magnesium strip and into the external circuit, which consists of wires leading into and out of a voltmeter, and towards the copper strip. As the electrons build up on the copper strip, Cu^{2+} ions from the electrolyte (copper(II) sulphate solution) are attracted towards them and combine to form copper atoms. So the copper strip gradually accumulates a new layer of copper, while the magnesium strip dissolves.

$$Cu^{2+} + 2e^- \rightarrow Cu$$

The movement of electrons in the metals and external circuit and of ions in the electrolyte solution, gives a continuous electric current in the circuit.

If we 'add' the two equations above, we can see that there is, overall, a reaction between copper ions and magnesium metal.

$$Mg \longrightarrow Mg^{2+} + 2e^-$$
$$Cu^{2+} + 2e^- \longrightarrow Cu$$
$$\overline{Mg + Cu^{2+} \longrightarrow Mg^{2+} + Cu}$$

This reaction happens when a piece of magnesium is placed in copper(II) sulphate solution. The magnesium is immediately coated with copper (figure 29). Copper ions are taking electrons directly from the magnesium surface. During this direct reaction the solution becomes warm. Here, chemical energy is converted into heat energy. In the electric cell, the chemical energy of the reaction is converted partly into heat energy and partly into electrical energy which drives electrons and ions in the circuit.

Figure 29 *Magnesium being coated with copper*

Improving simple cells

The simple cell does not work efficiently. After a short while, the voltage fails. You will see the magnesium surface gradually being coated with copper because there is a direct transfer of electrons from the magnesium atoms to the copper ions. This means that the electron flow round the circuit is reduced. And when there is no flow of electrons there is no current.

Simple cells are improved by separating the electrodes into two parts, known as **half-cells** (figure 30). Magnesium is now only in contact with its own ions so electrons are not 'lost' by direct reaction between magnesium and copper ions.

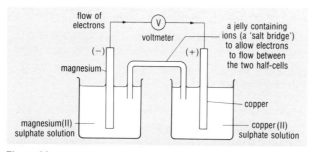

Figure 30

But just look at the size of the cell in figure 30. You can hardly imagine it sitting inside a torch or radio. It would be far too large and messy.

Buying a cell

When you buy a cell for a torch or radio, you will want something that is cheap, small, reliable and easy to use.

The metals used in the cell must therefore be:

◆ far enough apart in the electrochemical series to produce a useful voltage from the cell,

◆ not so reactive that they react too quickly with water in the electrolyte (this rules out lithium, sodium, potassium, calcium and even magnesium),

◆ cheap (this rules out silver and gold).

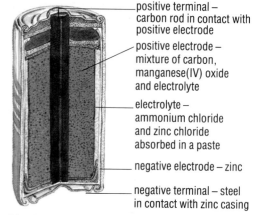

Figure 31 *A typical torch cell*

Figure 31 shows a typical cell suitable for a torch. The negative electrode is zinc and this is also the casing. It loses electrons to form zinc ions:

$$Zn \rightarrow Zn^{2+} + 2e^-$$

The electrons flow round the external circuit to the positive electrode. The positive electrode contains manganese(IV) oxide and carbon. The reaction at this electrode is very complicated but it can be simplified as

$$Mn^{4+} + e^- \rightarrow Mn^{3+}$$

The electrolyte of ammonium chloride and zinc chloride helps to conduct the electricity.

Many of the button cells you may use in watches and cameras also use zinc (in powdered form) for the negative electrode (figure 32).

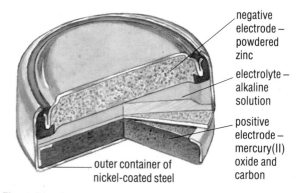

Figure 32 *A button cell*

Mercury(II) oxide is the positive electrode. The electrolyte is a concentrated solution of zinc oxide and potassium hydroxide and is alkaline.

The overall reaction is

$$Zn + HgO \rightarrow ZnO + Hg$$

and the voltage produced is 1.35 V.

Other oxides are also used instead of mercury oxide, for example, silver oxide (Ag_2O). These cells give a higher voltage (1.6 V) but are more expensive.

Using cells over and over again

The cells discussed so far have one disadvantage – once they have used up their chemical reactants they have to be thrown away. They are called **primary** cells.

However, the cells which make up the batteries in cars can be reused. They are called **secondary** cells. The best known is the lead-acid cell (figure 33).

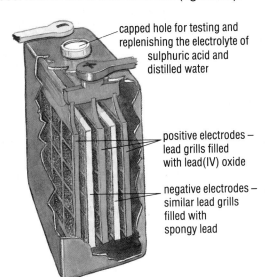

capped hole for testing and replenishing the electrolyte of sulphuric acid and distilled water

positive electrodes – lead grills filled with lead(IV) oxide

negative electrodes – similar lead grills filled with spongy lead

Figure 33   *A lead-acid cell*

The negative electrode is lead which loses electrons to become lead ions:

$$Pb \rightarrow Pb^{2+} + 2e^-$$

The positive electrode contains lead(IV) oxide which gains electrons. This reaction can be represented as:

$$Pb^{4+} + 2e^- \rightarrow Pb^{2+}$$

The electrolyte is sulphuric acid. It reacts with the Pb^{2+} ions to form lead(II) sulphate which is a white solid. It is insoluble in the acid solution. When the electrodes become covered with the white solid the cell goes flat.

The cell is recharged by passing an electric current through it backwards. This reverses the changes described above and the cell returns to its original state.

In theory, a lead-acid cell can be used and recharged for ever. In practice, the electrodes eventually crumble and the cell becomes useless. Lead-acid cells need looking after. Water can be lost from the electrolyte and the acid will then become too concentrated. Pure water – water that has been distilled – must be added to the cell when this happens (figure 34).

Figure 34   *Adding distilled water to a battery*

Taking it further

Lead-acid batteries are very heavy and large. They are not much use if you want to have a rechargeable cell in your radio or calculator. One rechargeable cell that is becoming popular is the nickel-cadmium cell.

The negative electrode is cadmium and releases electrons:

$$Cd \rightarrow Cd^{2+} + 2e^-$$

The positive electrode contains nickel(IV) oxide and gains electrons:

$$Ni^{4+} + 2e^- \rightarrow Ni^{2+}$$

Like the lead-acid cell, it can be recharged by passing an electric current through it backwards. The problem is that it does not keep its charge as well as many cells and it is also expensive.

6 How do we use electricity in chemistry?

Electricity is directly involved in the production of some of the most important chemicals, including metals. To understand how electricity is used you need to know what happens when it flows in a chemical compound. A suitable example is copper(II) chloride.

Electrolysis of copper(II) chloride

Copper(II) chloride in solid form will not conduct electricity. However, it does conduct well when it is dissolved in water as it forms ions in solution. When copper(II) chloride dissolves in water it releases copper ions ($Cu^{2+}(aq)$) and chloride ions ($Cl^-(aq)$). If carbon electrodes connected to a source of electricity are dipped into copper(II) chloride solution, the ions move through the solution towards the electrodes.

However there is more to it than that. You will observe that the electrode connected to the negative terminal of the source of electricity becomes coated with copper. Bubbles of a gas appear at the positive electrode. The gas, if collected, has a greenish colour and bleaches the litmus paper; it is chlorine.

Clearly, electricity has broken down copper(II) chloride into its elements, copper and chlorine. This is not the effect of heating, as copper(II) chloride does not break down to copper and chlorine even on strong heating. The name given to this decomposition of compounds using electricity is **electrolysis** (figure 35).

Figure 35 *Movement of ions a) before and b) during electrolysis of copper(II) chloride in solution*

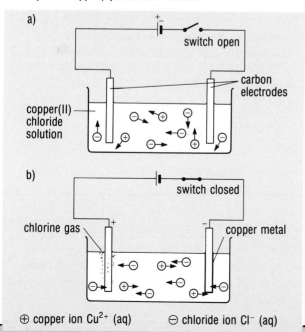

a) switch open
carbon electrodes
copper(II) chloride solution

b) switch closed
chlorine gas
copper metal

⊕ copper ion Cu^{2+} (aq)   ⊖ chloride ion Cl^- (aq)

What happens at the electrodes during the electrolysis of copper(II) chloride?

At the negative electrode:

positively charged copper ions ($Cu^{2+}(aq)$) are attracted to the negative charges. They gain electrons and become copper atoms which coat the electrode.

$$Cu^{2+}(aq) + 2e^- \rightarrow Cu(s)$$

At the positive electrode:

electrons from the negative chloride ions($Cl^-(aq)$) are attracted to the positive electrode, leaving chlorine atoms.

$$Cl^-(aq) \rightarrow Cl + e^-$$

Two chlorine atoms join to form a chlorine molecule and these rise in bubbles of gas.

$$Cl + Cl \rightarrow Cl_2(g)$$

Thus during electrolysis, electrons are removed from the negative electrode and are gained by the positive electrode. This keeps electrons moving in the electrodes and connecting wires of the circuit. In the electrolyte solution, copper ions ($Cu^{2+}(aq)$) keep moving towards the negative electrode and chloride ions ($Cl^-(aq)$) move towards the positive electrode. This is the electric current in the solution.

Electrolysis is the reverse of what happens inside electric cells (figure 36).

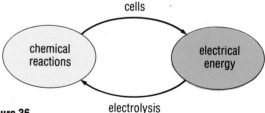

cells

chemical reactions   electrical energy

electrolysis

Figure 36

Electrolysis can only occur with compounds in a liquid state. Any chemical compound which forms ions when it melts or dissolves in water, is an electrolyte and can be electrolysed. This includes all acids, alkalis and salts.

During electrolysis:

metals or hydrogen gas are released at the negative electrode, for example:

$$Cu^{2+}(aq) + 2e^- \rightarrow Cu(s)$$
$$2H^+ (aq) + 2e^- \rightarrow H_2(g)$$

non-metals such as oxygen, chlorine or bromine are released at the positive electrode, for example:

$$2Br^-(aq) \rightarrow Br_2(g) + 2e^-$$
$$4OH^-(aq) \rightarrow O_2(g) + 2H_2O + 4e^-$$

Extraction of aluminium

Aluminium can only be produced cheaply if a large supply of low-cost electricity is available. It is extracted by the electrolysis of purified aluminium oxide (Al_2O_3), dissolved in molten cryolite (Na_3AlF_6).

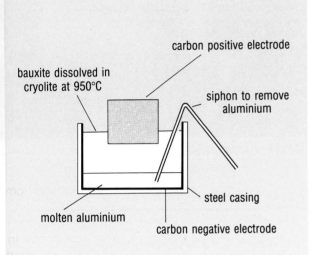

Figure 37

At the negative electrode:

$$2Al^{3+} + 6e^- \rightarrow 2Al$$

at the positive electrode:

$$3O^{2-} \rightarrow 3O + 6e^-$$

Oxygen atoms react with the carbon positive electrode to form carbon dioxide gas. After a while the carbon positive electrode must be replaced.

Making a tougher surface on aluminium (anodising)

Aluminium is a reactive metal protected by a strong layer of oxide on its surface. This layer can be made thicker and stronger by a process of electrolysis called **anodising**. The aluminium article is thoroughly cleaned, then it is made the positive electrode and dipped in dilute sulphuric acid. The negative electrode is another metal, such as lead.

At the positive electrode: hydroxide ions from water discharge to give oxygen.

$$2OH^- \rightarrow O + H_2O + 2e^-$$

The oxygen atoms react with the aluminium positive electrode to form more aluminium oxide. This oxide layer can hold dyes. Aluminium articles can be anodised to many different colours (figure 38).

Figure 38 *Anodised aluminium objects*

Chemicals from salt (sodium chloride)

Chlorine, hydrogen and sodium hydroxide are produced in very large quantities from the electrolysis of sodium chloride solution. This contains $Na^+(aq)$, $Cl^-(aq)$ and H_2O.

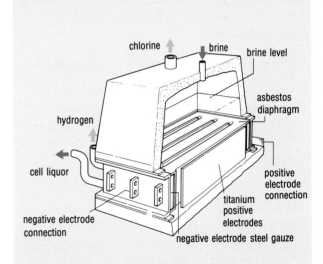

Figure 39

At the negative electrode:

$$2H_2O + 2e^- \rightarrow H_2(g) + 2OH^-(aq)$$

At the positive electrode:

$$2Cl^-(aq) \rightarrow Cl_2(g) + 2e^-$$

Sodium ions are not discharged. With the hydroxide ions, they form sodium hydroxide solution

$$Na^+(aq) \; OH^-(aq)$$

Chlorine, hydrogen and sodium hydroxide are used in making many other chemicals.

Producing pure copper by refining

Copper metal is first produced in crude form by the reduction of copper ores. Impure copper is then made the positive electrode. The negative electrode is a sheet of pure copper. The electrolyte is copper(II) sulphate ($CuSO_4$) in solution.

At the negative electrode: copper ions come out of solution by gaining electrons.

$$Cu^{2+}(aq) + 2e^- \rightarrow Cu(s)$$

At the positive electrode: copper ions go into the solution by losing electrons.

$$Cu(s) \rightarrow Cu^{2+}(aq) + 2e^-$$

The electrolyte solution does not change. It loses and gains copper ions at the same rate. The pure copper negative electrode gains mass and the impure copper positive electrode loses mass.

Figure 40 *A copper refinery*

Things to do

Things to try out

1   The first working electric battery was made by Alessandro Volta in 1800. He made his 'Voltaic Pile' using pairs of zinc and silver discs. Each pair was separated by a cloth soaked with salty water. Try to make a small voltaic pile using zinc and lead or copper. How large a voltage can you achieve?

Things to find out about

2   Find out, with dates, who
    a) defined one kind of electric charge as positive and another as negative,
    b) proposed the theory of ions to explain the conduction of electricity in liquid compounds,
    c) proposed the theory that electrons are minute 'particles' of negative charge,
    d) proposed the nuclear structure of atoms.

3   Sodium, potassium, calcium and magnesium are all extracted from their minerals by electrolysis. Answer the following questions for each metal.
    a) What is the main mineral from which extraction takes place?
    b) What has to happen to the mineral before it is electrolysed?
    c) What are the materials of the negative and positive electrodes?
    d) What are the reactions at the negative electrode?

e) What are the reactions at the positive electrode?
f) How is the metal extracted from the electrolysis apparatus?

4   Aluminium was so rare and precious a metal, over 150 years ago, that the French Emperor proudly owned a dinner service made from it. Yet aluminium compounds are among the most common on earth.
    a) Why was aluminium metal so rare?
    b) How was the metal for the Emperor's dinner service produced?
    c) What happened to make the production of aluminium much easier and more economical?
    d) Write a short account of the modern production of aluminium metal, starting from the mining of its ore.

5   What do the letters **e.p.n.s.** mean on metal cutlery or dishes? Explain what has happened to the metal items.

Points to discuss

6   When copper pipes carrying tap water were joined together with iron nuts, it was found after a few months that the iron nuts were badly corroded although the copper was unaffected. Suggest an explanation for this problem.

7   A well-known manufacturer of alkaline-manganese batteries claims that the batteries have many times the life of an ordinary battery. How would you test this claim scientifically?

8   Imagine a world where electricity cannot be produced from chemical energy. Electric cells do not exist. What would be the effect on human lives?

Questions to try

9   Write three short paragraphs which use all the words listed in a) to c) showing their correct meaning. The words can be in any order you choose.
    a) charge, positive, negative, conductor, current,
    b) atom, nucleus, protons, neutrons, electrons, ion, charge,
    c) electrolysis, electrolyte, electrodes, negative electrode, positive electrode, ions, electrons, charge, conductor, cell.

10 You are given the information in the table about six substances.

	Melting point/°C	Boiling point/°C	Conduction of electricity
A	−102	−24	high in solution
B	801	1412	only as melt
C	−52	76	none
D	540	2020	high as solid or liquid
E	24	137	none
F	2040	breaks down	only as melt

Choose one of **A** to **F** to answer each question below. Explain your reasons for your answer.
a) Which one is a metal?
b) Which one gives ions in solution?
c) Which one is an insoluble ionic compound?
d) Which one is a gas at room temperature?
e) Which one is a covalent liquid at room temperature?
f) Which one has close-packed molecules at room temperature?

11 What are the ions present in solutions of the following compounds? Give the symbol and the charge on the ions.

potassium carbonate   calcium hydrogencarbonate
lead nitrate   aluminium chloride
sodium sulphate   magnesium hydroxide

12 The Daniell cell (figure 41), invented by Professor J.F. Daniell in 1836, was one of the first cells to give a constant voltage over a long period of time.

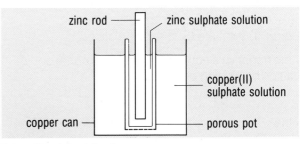

Figure 41   *A Daniell cell*

a) Which is the negative and which the positive electrode?
b) What reactions happen at the electrodes?
c) Why did Daniell keep the zinc rod in zinc sulphate solution in a porous pot?
d) Why does this cell give a steady voltage over a long time?
e) The cell will not work for ever, even if topped up with electrolytes. Give reasons for this.

13 In an experiment, you set up a number of cells as shown in figure 42. The copper half-cell is kept constant while the other half-cell is changed. For each change you recorded:
a) the voltage of the full cell,
b) the metal used as the negative electrode.

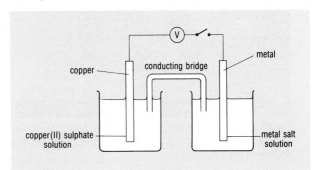

Figure 42

Cell	Voltage /V	Negative electrode
Copper–lead	0.5	lead
Copper–silver	0.3	copper
Copper–magnesium	1.5	magnesium
Copper–iron	0.5	iron
Copper–aluminium	0.9	aluminium
Copper–copper	0.0	–
Copper–nickel	0.3	nickel

Use the results in the table above to place the metals in a likely order of reactivity. Explain your reasons for your answer.

BURNING AND BONDING

Introducing burning and bonding

Figure 1 *Heat, light and sound energy are produced when chemical reactions take place in explosions like this controlled demolition of a condemned tower block.*

Burning is vital in everyday life. Most of the energy you use each day comes from the burning of fuels.

People burn fuels to keep warm. Most electricity is made by burning fuels in power stations. Your body gets its energy by using food as a fuel.

The burning of fuels gives out energy. All chemical reactions involve energy changes of some kind. Some reactions give out energy, some take energy in.

In this chapter you will see how

♦ chemistry helps explain the energy changes involved in burning and other reactions, and helps us control these reactions,
♦ chemistry explains why many chemical reactions need heating to get them started,
♦ chemistry helps us understand the properties of fire-resistant materials.

Figure 2 *The petrol being poured into this racing car is a convenient energy source. How does it release its energy? Why does it need heating to get it started?*

◀ **Figure 3** *How does this self-warming food pack work?*

◀ **Figure 4** *Gas is here being used directly as an energy source to heat food. What other methods of heating food do we use? What is their energy source?*

1 Fire-resistant materials

Fire-resistant building materials are important in making buildings safe. Fire-resistant materials need two properties in particular:

◆ They must not burn.
◆ They must have a high melting point.

These properties are decided by the material's chemical properties and structure.

Stone

Building stone is fire-resistant. Many types of building stone contain silicon(IV) oxide.

Silicon(IV) oxide has a giant structure of atoms joined by covalent bonds (see page 187). This is shown in figure 6.

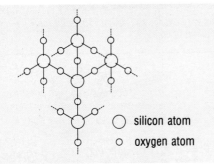

○ silicon atom
○ oxygen atom

Figure 6 *The structure of silicon(IV) oxide*

Substances with this type of structure always have high melting points because the atoms are all strongly bonded together. This makes them difficult to separate.

When a substance burns, its original bonds get broken and it forms new bonds to oxygen. Silicon(IV) oxide already has as many bonds to oxygen as it can make, so it does not burn.

Asbestos

Asbestos is fire-resistant and can be manufactured into hard, rigid shapes for use in building construction.

Asbestos forms long fibres and these can also be made into 'cloth', 'wool' and other useful materials. It does not burn and has a high melting point. For years it was used to make fire-resistant cloth and insulation.

There is more about the structure of asbestos in the chapter on Buildings.

◀ **Figure 5**
This wooden barn was completely destroyed by the fire. Modern farm buildings are built of fire-resistant materials.

Figure 7 *At the temperature of the fires on this burning oil tanker even the ship's metal plates will melt and burn.*

Asbestos substitutes

It has been discovered in recent years that asbestos can cause some kinds of cancer. Scientists have developed substitutes to take the place of asbestos. There is more about this in the chapter on Buildings.

Metals

Metals, like iron and aluminium, have quite high melting points. They do not normally burn but if they become very hot, they may melt and even catch fire. For this reason they are not ideal fire-resistant materials.

Choosing the right material Each of the following things needs to be made from fire-resistant material. In each case, suggest a suitable material:
1 a fire-resistant safe in a bank,
2 the inner lining of a furnace,
3 the door between a house and its garage,
4 a heat-resistant mat for school chemistry laboratories.

2 Burning liquid fuels safely

When fuels burn they combine with oxygen. Before this can happen, the fuel must **vaporize** so its vapour mixes with the air.

Some fuels vaporize more readily than others. Petrol vaporizes very easily. At room temperature there is enough vapour to **ignite** (catch fire) as soon as a match is held over the petrol (figure 9a). Paraffin vaporizes less easily, so it is harder to light (figure 9b). This makes it safer. To get paraffin to burn steadily, a wick or special burner is used.

Flash points

If you heat paraffin, it starts to give off more vapour. At a certain temperature (about 70°C), there is enough vapour to ignite when a lighted match is held near. This temperature is called the **flash point** of paraffin.

The flash point of a fuel is the *lowest* temperature at which there is enough vapour for the fuel to ignite.

The flash point of petrol is below *minus* 17°C. Petrol will ignite at any temperature above this, so petrol catches fire very easily.

Table 1 gives the flash points and boiling points of several flammable chemicals.

Figure 8 *Petrol vaporizes very easily, so it is easy to ignite, and great care must be taken when handling it.*

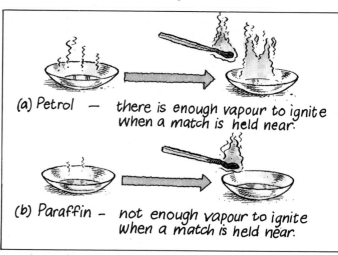

(a) Petrol — there is enough vapour to ignite when a match is held near.

(b) Paraffin — not enough vapour to ignite when a match is held near.

Figure 9 *Testing fuel vaporization at room temperature*

Table 1 *The flash points and boiling points of some flammable chemicals*

Substance	Flash point/°C	Boiling point/°C
benzene	−11	80
butane	−60	−1
ethanol (alcohol)	+13	78
ether	−43	35
glycerine	176	290
paraffin*	around 60	160–250
petrol*	below −17	40–70
white spirit*	around 32	150–200

*Paraffin, petrol and white spirit are mixtures of many different hydrocarbons. The composition of these mixtures varies, so their flash points and boiling points vary.

Use table 1 to answer these questions.

1 Is there a rough, general relationship between flash point and boiling point? If so, what is it?
2 Suggest a reason for any relationship between flash point and boiling point.
3 Which substance in table 1 is a gas at normal temperatures?
4 Which of the substances in table 1 would be the easiest to ignite?
5 Which of the substances in table 1 would be the safest to store, as far as fire risks are concerned?

How can you make paraffin burn better?

Because of its high flash point, paraffin is difficult to burn in a controlled way. You can improve the burning of paraffin in several ways. Here are two examples.

a) *Using a wick* (figure 10). This method is used in many paraffin lamps and heaters. The wick provides a steady flow of paraffin vapour to keep the flame going.

b) *By mixing paraffin vapour and air under pressure*. Figure 11 shows a 'Primus' stove. The paraffin fuel is pre-heated so it vapourizes. The vapour is mixed with compressed air. This mixture burns with a hot, blue flame.

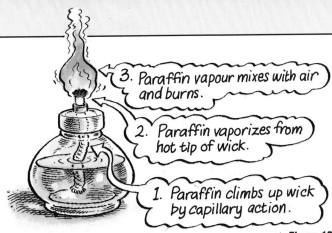

3. Paraffin vapour mixes with air and burns.

2. Paraffin vaporizes from hot tip of wick.

1. Paraffin climbs up wick by capillary action.

▲ **Figure 10**
A paraffin wick burner

Jet engines use the same general principle of burning a mixture of paraffin vapour and compressed air (figure 12). The hot exhaust gases are forced out of the back of the engine. This provides the thrust which pushes the plane through the air.

Figure 11 *The blue flame of a Primus stove*

6 Wick burners can also be used to burn petrol. Why is it difficult to burn petrol safely, and why do wick burners help?

7 In a wick burner, paraffin burns with a yellow flame. In a Primus stove, it burns with a blue flame. Suggest a reason for the difference. (Hint: think of the two types of bunsen burner flame.)

8 Jet engines cannot be used to propel space craft. Why not? What type of engines are used instead?

Figure 12
A cut-away view of an RB211 jet engine, which uses paraffin for fuel.

1 When a fuel burns, it reacts **exothermically** with oxygen (see also note 9 in this section).

2 The Fire Triangle (figure 13) shows the three things needed to make a fire. If one of these things is removed, the fire goes out.

Figure 13
The fire triangle

3 Fire extinguishers work by removing one or more of the three things in the Fire Triangle. Different types of extinguishers are needed for different types of fires.

4 Liquid fuels catch fire when there is enough of their vapour coming off to make a flammable mixture with air.

5 The flash point of a fuel is the lowest temperature at which there is enough vapour for the fuel to ignite.

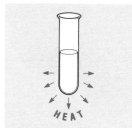

LOW FLASHPOINT
e.g. petrol, meths
• ignites easily
• dangerous to store

HIGH FLASHPOINT
e.g. paraffin
• difficult to ignite
• safe to store

Figure 14 *Some advantages and disadvantages of low and high flash-points in fuels*

6 Natural gas is an important domestic fuel. It contains mainly methane (CH_4). It is cheap and is easily piped to people's homes. It causes little pollution when it burns.

7 Mixtures of natural gas and air can be explosive. Gas burners are designed to allow a mixture of natural gas and air to burn in a controlled way.

8 When a chemical reaction happens, there is usually a temperature change.

9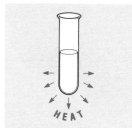

Figure 15 *Exothermic reactions give out heat*

Figure 16 *Endothermic reactions take in heat*

The energy changes in a reaction can be shown on an energy level diagram (see page 183).

10 Chemical reactions involve making and breaking bonds (figure 17). This results in energy changes.

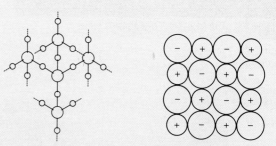

Breaking the bond takes energy **in**

Making the bond gives energy **out**

Figure 17

11 Many chemical reactions need heating to get them started. The energy needed to start a reaction is called its **activation energy**.

12 Activation energy is needed to break bonds, before new bonds can be made.

13 **Catalysts** speed up reactions but do not get used up. They speed up reactions because they lower the activation energy.

14 The uses of substances depend on their properties. These properties are governed by their structure.

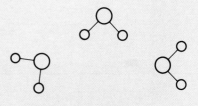

Giant structure of atoms with covalent bonds

Giant structure of ions held by ionic bonds

Simple molecules with covalent bonds between atoms

Figure 18 *Solid compounds have three types of structure*

1 What energy changes do you get with chemical reactions?

Reactions which give out heat are called **exothermic** reactions. Burning is exothermic. For example, when natural gas (methane) burns in air:

methane + oxygen → carbon + water + HEAT
dioxide

$$CH_4 + 2O_2 \rightarrow CO_2 + 2H_2O$$

Many reactions are exothermic. For example, whenever an alkali neutralizes an acid, heat is given out. Neutralization is exothermic.

Figure 19 *These campers are enjoying an exothermic reaction around their camp fire.*

But a few reactions *take in* heat. They are **endothermic**. During an endothermic reaction, the *substances* get cooler.

Figure 20 *Sherbet feels cool in your mouth when you eat it. The reaction of the sherbet with water is endothermic.*

Next time you eat some sherbet, see what happens to the temperature of your mouth. When sherbet mixes with water, there is an endothermic reaction. Your mouth feels cooler.

Sherbet is a mixture of sodium hydrogencarbonate and citric acid. When water is added, they react together like this:

HEAT + sodium + citric → sodium + carbon + water
hydrogencarbonate acid citrate dioxide

Notice that HEAT is on the left-hand side of this word equation. Heat gets taken in.

Using energy level diagrams

We can summarize these changes using **energy level diagrams** (figures 21 and 22).

Figure 21 is an energy level diagram for an **exothermic** reaction, like the burning of methane. The reactants have *more* energy than the products. As reactants turn to products, the extra energy is given out as heat.

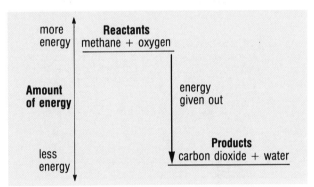

Figure 21 *Energy changes in an exothermic reaction*

Figure 22 is an energy level diagram for an **endothermic** reaction, like sodium hydrogencarbonate and citric acid. The reactants have *less* energy than the products. As reactants turn to products, they take energy in from the surroundings.

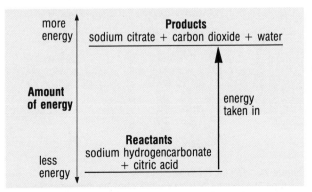

Figure 22 *Energy changes in an endothermic reaction*

Other forms of energy

Heat is only one form of energy. Light, electricity and sound are also forms of energy.

Sometimes chemical reactions give out or take in other forms of energy as well as heat. The photographs below (figures 23 and 24) give two examples of these energy changes.

Figure 23 *Some chemical reactions give out both light and heat energy. Can you think of a reaction that takes in light energy?*

Figure 24 *Electric cells like these use a chemical reaction to give out electrical energy. Can you think of a reaction that takes in electrical energy?*

2 Why do chemical reactions involve energy changes?

When methane reacts with oxygen, carbon dioxide and water are formed and heat is given out.

$$CH_4(g) + 2O_2(g) \rightarrow CO_2(g) + 2H_2O(g) + \text{HEAT}$$

Figure 25 shows the molecules involved in this reaction.

Figure 25

The atoms are joined together by strong chemical bonds called **covalent bonds**. During the reaction these bonds have to be broken, so each atom can form new bonds to different atoms.

Breaking a chemical bond needs energy. The energy is needed to pull the atoms apart. So before the reaction will start, energy needs to be put in to break the old bonds. This is called the **activation energy** (see page 185).

As the old bonds are broken, new ones are made.
Making chemical bonds releases energy.

In a chemical reaction there are two stages:

Stage 1 Breaking the old bonds, which *takes* energy.
Stage 2 Making the new bonds, which *gives out* energy.

The overall energy change for the reaction depends on the balance between these stages. If stage 1 involves more energy than stage 2, the overall reaction will take in energy – it will be **endothermic**. If stage 2 involves more energy than stage 1, the overall reaction will give out energy – it will be **exothermic**.

Taking it further

We can put some numbers to these energy changes.

Experiments show that when 1 litre of methane burns, about 30 kJ of heat are given out. The experimental results on the next page show where this 30 kJ comes from.

continued ▷

	Energy taken in/kJ	Energy given out/kJ
Stage 1 Breaking bonds		
Break 4 C – H bonds in methane	72	
Break 2 O = O bonds in oxygen	42	
Stage 2 Making bonds		
Make 2 C = O bonds in carbon dioxide		67
Make 4 O – H bonds in water		77
Totals	114	144

Overall amount of energy given out =

$$144 - 114 = 30kJ$$

3 Why do some reactions need heat to get them started?

Some reactions need heating to get them started. Natural gas, for example, will not start burning until it has been heated by a match or a spark. Some reactions need very little heat to get them started. For example, sodium hydrogencarbonate and citric acid (sherbet) react as soon as water is added.

Figure 26 *It is necessary to heat coal using burning paper and wood before it will start to burn. The reaction between coal and air has a high activation energy.*

Figure 27 *A spark provides the activation energy needed to start the reaction between petrol and air in a car engine.*

Activation energy

The energy needed to get a reaction started is called its **activation energy**. The energy is used to break bonds so that new bonds can start to form (figure 28).

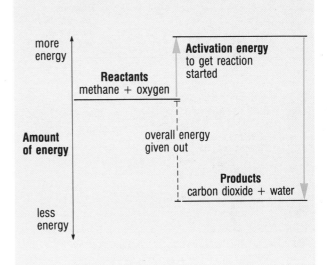

Figure 28 *This diagram shows the activation energy for methane burning in oxygen.*

If the reaction is exothermic, it will provide its own energy once it has started. You only need to light a gas burner once. Once it is going, it keeps going. But with an endothermic reaction, you have to keep supplying heat to keep the reaction going.

Catalysts and activation energy

Catalysts are substances which change the speed of a reaction without getting used up. Most catalysts *increase* the speed of reactions.

Catalysts work by lowering the activation energy of a reaction. With lower activation energy, it is easier for the reaction to get started and to keep going, so the reaction goes more quickly.

Taking it further

Many of the most important industrial uses of catalysis involve reactions between two gases on a solid. The solid is the catalyst and allows the gases to stick on its surface. This process is called **adsorption**. The gas molecules are adsorbed on to the surface of the solid.

This subject was introduced in the chapter on Minerals and now you can take a closer look at the way it happens.

The *Haber Process* is an example. In this process nitrogen reacts with hydrogen to form ammonia. The catalyst is iron (figure 29).

$$N_2(g) \; + \; 3H_2(g) \; \xrightarrow{\text{iron}} \; 2NH_3(g)$$

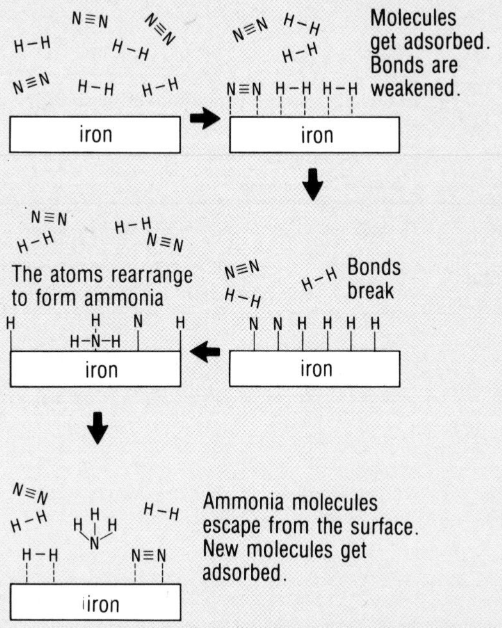

Figure 29 *How iron works as a catalyst in the Haber process*

The iron catalyst adsorbs nitrogen and hydrogen molecules onto its surface. If the surface is hot, the bonds in the molecules are weakened. They break to form atoms of nitrogen and hydrogen. These atoms form new bonds to each other to produce ammonia.

Without the iron catalyst, higher temperatures would be needed before the bonds started to break. The iron acts as a catalyst by lowering the activation energy. This means that the reaction can be carried out at a lower temperature, so the ammonia can be made more cheaply.

4 How are atoms and ions arranged in solid structures?

Different things are made from different materials. Concrete is used to make bridges and glass is used to make windows. You wouldn't use concrete for windows or glass for bridges, because they have the wrong properties. The materials for a particular use are chosen because they have the right properties for that use.

Figure 30 *Many different materials are used in making a car – metals, fabrics, plastics, leather, wood – and each is chosen because it has the correct properties for the job it has to do.*

The properties of a substance depend on its **structure** – the way its atoms, molecules and ions are held together.

Three types of structure are found in chemical compounds.

Giant structures of atoms joined by covalent bonds
In this particular type of structure, the atoms build up into a giant, regular network, with every atom joined to several others. The giant structure continues indefinitely.

Silicon(IV) oxide has this type of structure (see figure 6 on page 179). Sand is mainly silicon(IV) oxide, and so are many kinds of rock.

Because the atoms are all strongly bonded together, they are difficult to separate. This makes the substance hard, with a high melting point. The substance will not dissolve in water. There are no charged particles, like electrons or ions, which are free to move around the structure. This means the substance does not conduct electricity.

Many important building materials, such as brick, stone and concrete, have this type of structure.

Simple molecules with covalent bonds between atoms within the molecules

This type of structure also contains atoms joined by covalent bonds. But unlike the giant covalent structure, each atom is only bonded to a small number of other atoms. The structure does *not* go on indefinitely: it contains small, simple molecules. Water has this type of structure (figure 31).

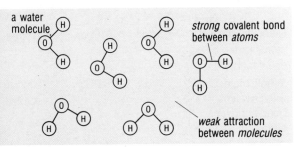

Figure 31 *The structure of water*

The atoms are strongly bonded together within each molecule. But there are only weak attractions *between* molecules, so the molecules are easily separated. This makes the substance easy to melt and boil. In fact, substances with this kind of structure are often liquids or gases.

Methane, carbon dioxide, alcohol and wax all have this type of simple molecular structure.

Simple molecular substances *may* dissolve in water, because their molecules can be easily separated and can fit between water molecules. They do not conduct electricity, because their particles have no electric charge to carry the current.

Figure 33 *The Dead Sea – a place of sand, water and salt. These three compounds all have very different structures and yet they occur together. Why does the salt dissolve in the water but not the sand? Why are the sand and salt solid but not the water?*

Giant structures of ions held by ionic bonds

This type of structure contains ions rather than neutral atoms. The ions are arranged in a regular giant structure, with oppositely-charged ions attracting one another. The giant structure continues indefinitely.

Sodium chloride (ordinary table salt) has this type of structure, containing Na^+ and Cl^- ions (figure 32).

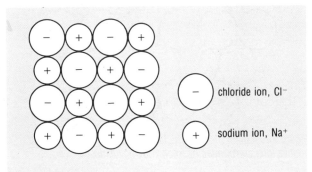

Figure 32 *The structure of sodium chloride. The diagram shows just one layer of a giant structure.*

Because the ions are all strongly bonded together, they are difficult to separate. This makes ionic substances hard, with high melting points. But the charged ions are attracted to the charges on polar water molecules, so this type of substance usually dissolves in water. When ionic substances are molten or dissolved, the ions are free to move around. This means they can carry an electric current, so the substance conducts electricity.

All compounds of metals with non-metals are ionic, for example, calcium oxide (lime) and copper(II) sulphate.

Other types of structure

The three types of structure mentioned above account for all compounds. But *metals* have a fourth type of structure, the **giant metallic structure**.

Metals tend to lose electrons easily. When they lose electrons, they form positive ions. In a giant metallic structure, these positive ions are arranged in a regular way (figure 34).

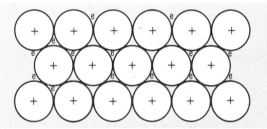

Figure 34 *A giant metallic stucture – it has regularly arranged postive ions surrounded by freely moving electrons*

The electrons, being very small, are able to move freely between the positive ions. So a giant metallic structure has regularly-arranged positive ions surrounded by freely-moving electrons. These freely moving electrons enable the metal to conduct electricity.

Table 2 summarizes the properties of compounds with different types of structure.

Table 2

Type of structure	Melting point	Electrical conductivity	Solubility in water
Giant structure of atoms joined by covalent bonds	High	Poor	Insoluble
Simple molecules with covalent bonds between atoms	Low	Poor	May be soluble
Giant structure of ions	High	Good when molten or dissolved	Usually soluble
Giant metallic	Usually high	Good	Insoluble

Things to do

Things to try out

1 *A survey of ignition methods*
   Have a good look around your home. How many different ways of igniting things can you see? How does each method work? Remember to include things like cookers and gas boilers. Also include car engines and other things to be found in the garage, garden and so on.

2 *Fire risks in your home*
   Look around your home. What do you think are the major fire risks? In other words, what are the places and situations where a fire is most likely to break out? In each case, what would be the best action to take to get the fire under control?

Things to find out

3 a) Find out the difference between a petrol engine and a diesel engine.
   b) What are the important differences between petrol and diesel fuel?

4 Go to a local petrol station. Ask them about their fire-prevention routine.
   a) What are the major fire risks?
   b) How do they keep these risks to a minimum?
   c) What is their standard routine if a fire *does* break out?
   d) What kinds of fire-extinguishers do they have at the petrol station?

5 Some camping shops sell 'hand-warmers'. Each hand-warmer consists of a small packet made from porous fabric. The packet contains a dark grey powder. When you buy the hand-warmer, the packet is contained in an outer, airtight plastic bag. The following directions are given for using the hand-warmer.

Directions
Open the outer plastic bag. Remove the inner packet. Shake it several times. Hold the packet in your hand. It will keep at a comfortable 60°C for several hours.

Ingredients
Powdered iron, water (adsorbed on cellulose), salt.

   a) Try to decide how the hand-warmer works.
   b) What experiments could you do to check whether your explanation is correct?

6 Find examples of catalysts that are used in industrial processes. In each case, write down the name of the catalyst and the chemical reaction which it speeds up.

Points to discuss

7 How do forest fires start? How could forest fires have started before there were humans on earth? Do you think forest fires do more or less damage now than in the days before humans?

8 Every year people are killed because of home fires involving furnishing materials. Often they are killed by poisonous fumes given off by the foam used to fill armchairs and sofas. Could these deaths be avoided? If so, how? What are the problems in carrying out your suggestions?

9 Would it be possible to make a home *completely* safe from fire risks? What would be the biggest problems you would have to solve?

10 It has been said that all the energy used by humans comes from the making and breaking of chemical bonds. Is this true? If not, try to think of some exceptions.

Questions to try

11 Fill in the missing words or groups of words in the following. Each word or group of words is used only once.

The missing words or groups of words are: activation energy, break, endothermic, exothermic, given out, greater, less, taken in.

Some chemical reactions give out heat. They are described as ___ (a) ___ reactions. Some reactions take in heat. They are described as ___ (b) ___ reactions. All chemical reactions involve breaking and making chemical bonds. When a bond is broken, energy is ___ (c) ___. When a bond is made, energy is ___ (d) ___ . Before a reaction can start energy must be supplied to start it off. This energy is needed to ___ (e) ___ the existing bonds. This 'starting off' energy is called the ___ (f) ___ . Once the old bonds have been broken, new ones can form. This releases energy. If the energy needed to break the old bonds is ___ (g) ___ than the energy released by making the new bonds, the reaction is exothermic. But if the energy needed to break the old bonds is ___ (h) ___ than the energy released by making the new bonds, the reaction is endothermic.

12 Below you will see a list of reactions. For each reaction:
a) Say whether you think the reaction is exothermic or endothermic.
b) Draw an energy level diagram like the ones in figures 21 and 22 on page 183.

(i) The reaction of hydrogen with oxygen to form water.

hydrogen + oxygen → water

(ii) The decomposition of potassium chlorate by heat, forming potassium chloride and oxygen.

potassium → potassium + oxygen
chlorate   chloride

(iii) The burning of sulphur in air to form sulphur dioxide.

sulphur + oxygen → sulphur dioxide

13 This chapter uses the idea of breaking and making chemical bonds to explain the energy changes during chemical reactions. Use this idea to explain in your own words why
a) fuels need heating before they will catch fire,
b) once they have caught fire, fuels go on burning without further heating.

14 When hydrogen reacts with oxygen the following reaction takes place.

hydrogen + oxygen → water
$2H_2$ + O_2 → $2H_2O$

a) What bonds are *broken* during this reaction?
b) What new bonds are *made* during this reaction?
c) When 1 litre of hydrogen reacts with oxygen, about 5000 J of energy are released. Explain where this energy comes from.
d) Hydrogen normally only reacts with oxygen if it is heated. However, the two gases will react at room temperture if they are in contact with a gauze made of platinum metal. Explain.

15 Below you will see descriptions of the four different types of structure which are commonly found in solids.
a) Giant structure of atoms joined by covalent bonds.
b) Simple molecules with covalent bonds between atoms.
c) Giant structure of ions.
d) Giant metallic structure.

For each type of structure, give **three** examples of substances with that structure which could be found in your home.

Introducing fighting disease

The fight against disease goes on both outside and inside your body.

To fight disease *outside* your body you have to be sure that the air you breathe, the water you drink and the food you eat is safe. Also you have to keep the outside of your body clean.

Inside your body you have to try to keep the chemical reactions which are going on all the time working properly. Sometimes the balance of chemicals in your body is not quite right so you may be able to take medicines to adjust this balance. Sometimes the chemical reactions in your body are interfered with by micro-organisms and you can take medicines to fight these micro-organisms.

In this chapter you will see how

◆ germicides work and their role in public health,
◆ chemical reactions take place inside your body and the part played by enzymes in these reactions,
◆ the basic ideas about the action of drugs are related to their structures.

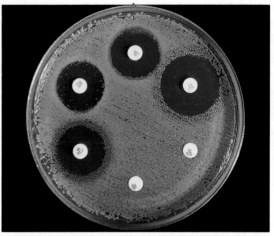

◄ Figure 1 This experiment shows how effective six antibiotics are against the bacteria E. coli. Two antibiotics have no obvious effect on the growth of the bacteria.

◄ Figure 2 Once a disease has taken hold, medicines may help to make you better. What medicines have you taken? In what form were they? How did they work?

Figure 3 Infusion pumps measure and deliver a regular drug dosage and allow patients to be treated over long periods away from hospitals. ▼

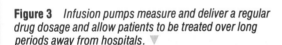

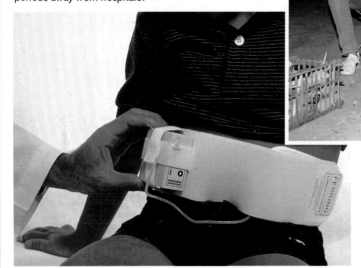

▲ Figure 4 How does boiling water make it safe to drink? What else could you do to make sure it is safe?

1 Clinical chemistry

Disease is caused by or results in changes in the chemical reactions which occur in your body. Some of these changes can be detected by analysing the body fluids – in particular blood and urine.

Some tests measure the concentration of a particular substance. For example, if a person is suffering from anaemia, the concentration of iron in the blood will be less than it should be. This is fairly common during pregnancy as the baby uses a lot of the mother's iron. Blood can be quickly tested by putting a drop into a test tube of liquid and watching how fast it falls through the liquid. This is using the blood's density as a quick guide to its iron content. If the nurse suspects that the person might be anaemic, then another sample of blood will be sent to the hospital for more accurate analysis.

If the blood test shows that the person is short of iron, then a course of 'iron tablets' is prescribed. The so called 'iron tablets' (figure 7) contain iron in the form of a compound which can be absorbed by the body. The active ingredient in the tablets is sometimes iron(II) sulphate. In your body, iron is combined with haemoglobin which is the part of the blood cells which transport oxygen around your body.

Sometimes a test is to see if a substance is present in *higher* concentrations than would be normal. An important example of this is sugar in urine. Glucose is the sugar which

◀ **Figure 5**
A clinical chemist at work in a hospital laboratory

Figure 6
Taking a blood sample for initial testing at a blood-donor clinic ▶

would be present in the urine of a diabetic (see page 50). It can be tested for in exactly the same way that it can be tested for in foods, i.e. by warming with Benedict's solution.

An important advance in testing for sugar in urine is the use of diagnostic test strips. These are small plastic sticks with the test chemicals absorbed in a porous material on one end. When the stick is dipped into a solution containing sugar, the test chemicals change colour within 15 seconds. The colour can then be matched against a colour chart to give an approximate indication of the sugar concentration (figure 8). The test chemicals on the stick used for sugar testing are a mixture of enzymes and colouring agents.

Figure 7   *'Iron' tablets come in various forms* ▼

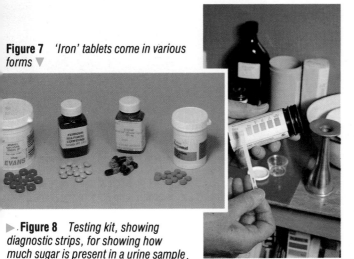

▶ **Figure 8**   *Testing kit, showing diagnostic strips, for showing how much sugar is present in a urine sample.*

1 Find the following words in the text and explain what each means using examples from the text to illustrate your answers: body fluids, test, analysis, reducing, diagnostic.
2 In what ways do you think diagnostic test strips have been helpful to people suffering from diabetes?
3 Imagine you are a clinical chemist working in a hospital. Write a diary describing what you might do in a typical day at work.

2 Killing the germs – chemotherapy

Chemotherapy is the use of chemicals to attack harmful micro-organisms in the body.

Natural products have been used for centuries to treat diseases. For example, there is an account in 1619 of the wife of the Spanish Governor of Peru being successfully treated for malaria with an extract from the bark of a chinona tree (figure 9). The active chemical in the bark was quinine. It was not until 1820, over 200 years later, that quinine was extracted from the bark by two chemists working in France.

The most dramatic step forward in chemotherapy was the development of **antibiotics**. In 1928, Alexander Fleming was doing research in St Mary's Hospital, London. He was growing bacteria in petri dishes and left some dishes on his bench when he went on holiday. When he returned he noticed that there was a mould growing in one dish as well as bacteria (figure 10). But more importantly he also noticed that the bacteria were **not** growing near the mould. This was the penicillin mould. A substance called **penicillin** was eventually separated from it and this was found to kill bacteria.

Penicillin was the first antibiotic to be identified. An antibiotic is a substance extracted from a living micro-organism which will attack other harmful micro-organisms in the body.

Figure 9 *The chinona tree from which quinine is extracted for use in the treatment of malaria.*

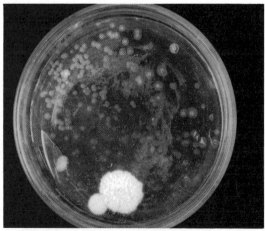

▲ **Figure 10** *The culture plate on which Sir Alexander Fleming noticed the penicillum mould growing.*

◄ **Figure 11** *Sir Alexander Fleming in his laboratory. Had he not been so observant on his return from holiday he might not have discovered penicillin.*

Figure 12
Large-scale production of penicillin

It is thought that antibiotics work in a number of different ways. It is clear that some stop the bacteria growing by occupying active sites in important molecules in the bacteria. (Active sites – 'holes' in enzymes – are discussed on page 197.) Others appear to destroy the cell walls of the bacteria. They are all **systemic antibacterial** substances as they reach the bacteria by circulating in the blood stream.

An important consequence of using antibiotics which was not realised at first is that sometimes resistant strains of bacteria develop. This happens when an antibiotic kills off a particular type of bacteria but has no effect on another closely related type of bacteria which is also present.

This second type might grow more rapidly when the first one has been killed and so the antibiotic is encouraging the growth of resistant bacteria. This is why doctors prescribe a carefully calculated course of antibiotics and why they will only use them when absolutely necessary. It is also why if you are given a course of antibiotics you should take the whole course as directed.

Fleming did not find any evidence that penicillin would attack bacteria in the body. So it was really not until 1940 that the full importance of penicillin was recognised. This was during World War II when it was realised that if it could be produced in sufficient quantities it would save the lives of many who would otherwise die of infected wounds.

Some antibiotics have been discovered in what might appear unlikely places, for example, one is reported to have been first found in a damp patch on a wall in Paris. Others were found after painstaking investigations. In America, the investigation of 10 000 soil cultures led to the discovery of **streptomycin** in the 1940s. This antibiotic had a dramatic effect on the treatment of tuberculosis which until that time killed many people. This antibiotic plus preventative and screening procedures has almost wiped out tuberculosis in this country.

1 Find the words chemotherapy, antibiotic, systemic and technological in the text and explain what you think each of them means.
2 Suggest why it is recommended that some antibiotics should be taken at the time of day when your stomach is least full.
3 Suggest why the use of antibiotics in farm animals must be carefully controlled.

Figure 13 *Killing off one type of bacteria …*

Figure 14 *… can encourage the growth of more resistant bacteria.*

1   Fighting disease involves caring for

the external environment:
- clean air,
- clean water,
- clean food,
- clean bodies,
- clean environment.

the internal environment:
- adjusting the chemical balance in our bodies,
- fighting micro-organisms in our bodies.

2   Germicides are substances used to kill germs *outside* the body. Germs are harmful micro-organisms such as bacteria.

Figure 15

Disinfectants are more concentrated and less pure than antiseptics. Disinfectants in particular should be kept out of the reach of young children.

3   The most common disinfectant is chlorine. The gas is used to kill germs in our water supply and in swimming pools. The pH of swimming pool water which has been treated with chlorine must be carefully controlled. If the pH becomes too low (more acidic) chlorine gas is given off. If it becomes too high (alkaline) the germicidal action is reduced.

Household disinfectants and some liquid antiseptics are solutions of a compound of chlorine called sodium hypochlorite. This is made by reacting chlorine with dilute sodium hydroxide.

4   A wide range of drugs is produced to help to fight disease *inside* the body. These may fight micro-organisms or adjust the chemical balance in the body.

5   A large proportion of your body is water. Many complex chemical reactions are occurring in your body all the time. All of these reactions occur between substances dissolved in water. Your health depends on these reactions working properly.

6   All the reactions occur at the temperature of the body. Many of the reactions are able to occur at this relatively low temperature because your body contains special chemicals which speed up the reactions. These substances are called **enzymes**.

Substances which speed up reactions but are not used up themselves are called catalysts. Enzymes are *biological* catalysts.

Figure 16 *Enzymes catalyse reactions inside you.*

7   An example of a chemical reaction which takes place inside your body is the conversion of starch to glucose.

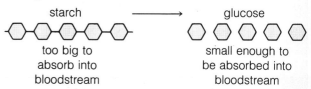

starch → glucose

too big to absorb into bloodstream

small enough to be absorbed into bloodstream

This reaction is catalysed by the enzyme amylase which is found in saliva.

8   Enzymes are fussy. Each one catalyses a particular reaction. Also their effectiveness is influenced by the pH of the solution and they work better within certain small temperature ranges.

Enzymes are destroyed by high temperatures (usually above 45°C). An enzyme has a particular area, called an 'active site', where the reaction it catalyses takes place.

9   When harmful micro-organisms enter the body they can interfere with the chemical reactions of the body by stopping the enzyme action.

germs   germs

A + B + enzyme → products

germs   germs

But micro-organisms also contain enzymes and so it is sometimes possible to attack them by taking antibiotics which interfere with the action of their enzymes.

antibiotic   antibiotic

germ   germ

A + B + enzyme → products

germ   germ

antibiotic   antibiotic

1 What are germicides?

Insecticides kill insects, herbicides kill weeds and **germicides** kill **germs**.

Germs are **micro-organisms**, which means they are very small living things. They are too small to be seen without a microscope. Some germs are called **bacteria**. Not all bacteria are harmful. For example, some bacteria in the intestine produce vitamin K which helps your blood to clot.

Other bacteria can be harmful because they cause infections or diseases. Germicides are chemicals used to destroy harmful bacteria which are outside the body. Those which are used on living things are called **antiseptics** and those which are used in water or on objects are called **disinfectants**.

The chemically active part of disinfectants and antiseptics can be the same substance, but in the case of the antiseptic the concentration and purity of the substance must be more carefully controlled.

A lot of germicides work because they are oxidising agents which means that they supply oxygen which kills the bacteria.

Figure 17 *a) A mild disinfectant is used to kill the germs on a baby's bottle.*
b) Chlorine is used as a disinfectant to kill germs in swimming pools and in drinking water.
c) An antiseptic is used to clean a wound to prevent infection.

2 How can some of the properties of chlorine be explained?

Chlorine is a poisonous pale green gas (figure 18a).

It dissolves in water forming a pale green solution called **chlorine water** (figure 18b).

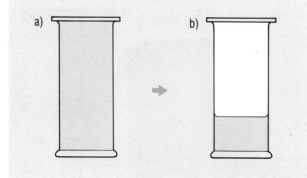

Figure 18

Chlorine water will:

◆ turn moist blue litmus paper red and then bleach it – this is used as a **test** for chlorine,
◆ give off oxygen gas when left in sunlight,
◆ react with a carbonate giving off bubbles of carbon dioxide.

These properties suggest that chlorine must react with water when it dissolves in it. A mixture of two acids is formed when chlorine is bubbled into water:

chlorine + water → hydrochloric + hypochlorous
$\quad\quad\quad\quad\quad\quad\quad\quad$ acid $\quad\quad\quad$ acid
$\quad Cl_2 \quad + H_2O \rightarrow \quad HCl \quad + \quad HOCl$

The fact that the solution contains acids explains why it turns blue litmus red and why it reacts with carbonates to form carbon dioxide.

One of the acids in chlorine water is hypochlorous acid which is unstable and easily decomposes to form oxygen gas:

hypochlorous → hydrochloric + oxygen
$\quad$ acid $\quad\quad\quad\quad$ acid
$\quad 2HOCl \quad \rightarrow \quad 2HCl \quad + \quad O_2$

It is the hypochlorous acid in the chlorine water which enables it to bleach some dyes and to act as a disinfectant.

In a swimming pool it is convenient to use chlorine gas from a cylinder to disinfect the water. In our homes this would be both dangerous and inconvenient and so the chlorine is converted to another compound by reacting it with sodium hydroxide solution.

Sodium hydroxide is an alkali and so it will react with an acid to form a salt plus water. Chlorine water contains a mixture of two acids and so a mixture of two salts will be formed:

chlorine + sodium → sodium + sodium + water
$\quad\quad\quad$ hydroxide $\quad$ chloride $\quad$ hypochlorite
$\quad Cl_2 \quad + \; 2NaOH \rightarrow \; NaCl \quad + \quad NaOCl \quad + H_2O$

Sodium hypochlorite is an oxidising agent and it is the most commonly used germicide. If you inspect the labels on bottles of household disinfectant you will probably see that sodium hypochlorite is the main ingredient.

3 What is happening inside our bodies?

All animals including humans stay alive by taking food, water and oxygen into their bodies.

Food is a *mixture* of *chemicals*.
Water is a *chemical*.
Oxygen is a *chemical*.

Our bodies are made up entirely of a variety of chemicals (figure 19).

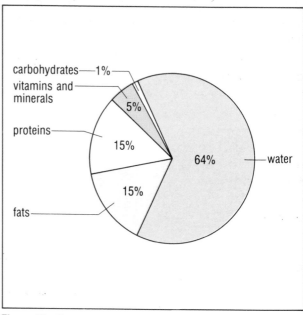

Figure 19 *Percentage composition, by mass, of a human body*

Your body works by means of many chemical reactions occurring inside it. Most of the reactions convert food chemicals into other chemicals which are *needed* by your body and into other waste chemicals which are *excreted* from your body. When you are healthy these reactions are working properly. Diseases interfere with some of the reactions and make you feel unwell.

4 What do enzymes do?

You grow and keep alive by chemical reactions occurring inside your body. All these reactions must work at the temperature of your body. This is the case for all living things – their chemical reactions work at the temperature of the plant or animal.

The temperature of your body is around 37°C. This means that the many complex reactions which occur in your body work at this temperature. Often it is more difficult for the same reactions to occur outside your body. For example, starch in food is converted to sugar as the food is digested. Outside the body starch needs to be boiled (100°C) with dilute hydrochloric acid to change it to sugar but in your body this reaction occurs at 37°C. The secret is that your body contains **enzymes**. These are biological catalysts which enable the reactions to occur at the temperature of your body.

Not only are enzymes very efficient catalysts but they are also very specific. This means that a particular enzyme is needed to catalyse a particular reaction.

Enzyme molecules have particular three-dimensional shapes and chemists have developed a theory of how they work as catalysts which depends on these shapes.

The theory is that each catalyst has a 'hole' in its shape (figure 20a).

The molecules involved in the reaction are attracted to holes and fit into them rather like a key fitting exactly into a lock (figure 20b).

While they are in the holes the reaction occurs and then the products are released from the holes (figure 20c).

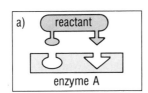

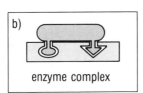

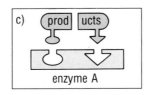

Figure 20 *How enzymes are thought to work as catalysts* ▶

When the efficiency of enzymes acting as catalysts is investigated it is found that they work better within a small temperature range and at particular pH values. Under different conditions the shape of the hole sometimes changes and the molecules in the reaction will no longer fit into them. This means the enzyme will no longer act as a catalyst.

Yeast contains enzymes which catalyse the conversion of sugar into alcohol. When brewing lager the mixture must be kept at a temperature around 10°C and at 20°C to produce beer.

If the temperature is raised above 50°C, the enzymes in the yeast are destroyed and the fermentation will stop.

Figure 21 *This sophisticated control system ensures that temperature fluctuation is kept to a minimum when a beer is brewed.*

5 Enzyme blockers

Enzymes are very sensitive to temperature and pH change. Sometimes they are very sensitive to the presence of other substances.

If bacteria produce a substance which attaches itself strongly to the hole in the enzyme it will stop the enzyme acting as a catalyst. This is the way that some bacteria have harmful effects on how the reactions in the body work and so make you feel ill.

The action of some drugs which kill bacteria can be explained in a similar way. Bacteria contain their own enzymes which are necessary to catalyse the reactions which result in the bacteria growing. If a drug is used which blocks the holes in the enzymes (figure 22) the bacteria will not be able to grow. In this way the drug will have removed the bacteria and will make you feel better.

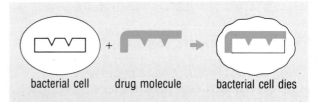

Figure 22 *How enzyme blockers work*

Things to try out

1 Add some household bleach to a small portion of washable ink. Use your observations to devise a way of comparing the strengths of different household bleaches.

2 Using brewers yeast and sugar solution carry out a series of experiments which will enable you to investigate over what temperature range the enzyme in the yeast works best.

Things to find out

3 Enzymes in yeast catalyse the fermentation reaction which converts sugars to alcohol. Different crops provide the sugars for different alcoholic drinks.
   a) Find out which crops are used to produce wine, cider and beer.
   b) Draw a flow diagram to represent the various stages involved in converting one of these crops into a drink.

4 Since the late 1940s cases of tuberculosis have been successfully treated by antibiotics. But at that time the disease was so common that there were national campaigns to prevent the disease spreading. Using other books and perhaps by questioning people who can remember those times find out what happened during these campaigns.

Things to write about

5 The headline in figure 23 appeared in a national newspaper.

Drug-resistant germ blamed for deaths
Deadly 'super bug' spreads through wards

Figure 23

Write an article (about half a page) which this headline could have been meant for. The article should not only report the events of the case but also inform the reader of the scientific principles involved and identify the issues.

6 Imagine you were the first scientist to notice that bacteria would not grow near a mould which had developed in a petri dish. Write a report of this observation so that other scientists could learn about your discovery. The report should explain what you did, what you observed, what explanation you are suggesting for the observation and what the possible implications are for fighting disease.

Points to discuss

7 In 1985-86 there was a whooping cough epidemic and about 35 000 children caught the disease, which in severe cases can have serious effects. A vaccine which prevents children catching whooping cough is available. A research study has shown that the vaccine has serious side effects in 1 in 100 000 cases. Do you think young children should be given the whooping cough vaccine?

8 Imagine you have a sore throat but otherwise you are healthy. Your doctor tells you that if you took antibiotics it would go away quickly but she will not give them to you because she believes they should only be used when absolutely necessary. Discuss whether or not you think the doctor has done the correct thing.

Questions to try

Questions 9–14

A antibiotic	B disinfectant	C bacteria
	D enzyme	E antiseptic

From the names, **A** to **E**, choose the one which

9 is a drug,

10 catalyses reactions in the body,

11 is alive,

12 is a germicide which can be used on the skin,

13 can cause infections,

14 is a germicide which should not be used on the skin.

15 Draw and complete table 1 which is concerned with important reactions involving catalysts.

Table 1

Reactants	Catalyst	Product	Use of product
nitrogen + hydrogen			for making fertilizers
sulphur dioxide + oxygen		sulphur trioxide	
sugar solution	enzymes in yeast		

16 The proportion of people dying from particular causes has changed a lot over the last 130 years. Table 2 gives the approximate percentages of deaths from four particular causes in 1850, 1910 and 1970.

Table 2

	Infectious diseases	Tuberculosis	Heart disease	Cancer	Other causes
1850	28	17	3	2	50
1910	15	11	9	8	57
1970	1	1	32	32	34

During this time the average life expectancy (the average number of years a person born in a particular year could expect to live) has also changed. Table 3 provides some data on this.

Table 3

Year of birth	Life expectancy / years
1850	40
1900	46
1930	58
1970	70

Display the data in table 2 as pie charts and the data in table 3 as a graph.

a) Write a short paragraph describing the changes in the proportion of people dying from particular diseases.

b) Explain how water treatment has contributed to the decline in infectious diseases.

c) Chlorine is used as a disinfectant for the treatment of swimming pool water.
   (i) What is formed when chlorine dissolves in water?
   (ii) Which substance in chlorine water is responsible for its sterilising action?
   (iii) Explain why it is important that the pH of a swimming pool should not fall below 7.

d) Antibiotics and screening have contributed to the changes in deaths due to tuberculosis.
   (i) What are antibiotics?
   (ii) What is medical screening?

e) Use the pie charts and the graph you have drawn to help you suggest an explanation for why there have been changes in the proportions of people dying from heart disease and cancer.

Introducing energy today and tomorrow

Everyday people use energy in thousands of different ways. For example, it is used to warm homes, cook meals and light the streets. Although a lot of energy is used in our homes, just as much is used in industry. Energy is needed to extract minerals, to generate electricity and to turn things like clay and iron ore into useful products like bricks and steel.

At one time, almost all the energy used in homes and industries came from fossil fuels – coal, oil and natural gas.

It is possible that oil and natural gas will begin to run out in your lifetime. Because of this, scientists and engineers are developing more and more ways of using alternative energy sources. Two alternatives are shown in figures 2 and 3.

◄ **Figure 1**
The chemical energy in coal is being used to turn things like iron ore into steel. The first part of the process takes place in a blast furnace like this one. What is happening to the iron ore in the blast furnace?

◄ **Figure 2** In a hydroelectric power station the energy from falling water is used to generate electricity. Where in the UK would you expect to find power stations like this? Give reasons for your answer.

In this chapter you will see

◆ how fossil fuels are formed and plants trap energy,
◆ why it is important to begin to use alternative energy sources in place of fossil fuels,
◆ what advantages uranium has over fossil fuels as a fuel for power stations,
◆ what the drawbacks are in using uranium,
◆ what other alternative energy sources can be used in place of fossil fuels.

Figure 3 This nuclear power station is at Sizewell in Suffolk. In a nuclear reactor, a controlled reaction occurs in the nuclei of uranium atoms. Energy is released at a steady rate and is used to generate electricity. ▶

1 Energy from underground rocks – geothermal energy

In some parts of the world, such as New Zealand, the Philippines, California and Iceland, there is hot water in cracks and pores in the rocks beneath the ground. The water in the rocks has been heated by volcanic activity. This hot water can be piped to the surface through drill holes. It can then be used to heat homes and other buildings. It can also be boiled, producing steam, which is then used to drive turbines in power stations and produce electricity.

Unfortunately, there is no useful hot water in the rocks beneath Britain. However, the temperature of rocks does increase with depth. In Cornwall, scientists and engineers are trying to extract the heat from hot, 'dry' rocks. This is called **hot, dry rock geothermal energy**.

The project is situated in a disused granite quarry at Rosemanowes near Penryn (figure 6). At Rosemanowes, the granite rock extends from the surface to a depth of 14 km where the temperature is about 400°C. To generate electricity, the temperature of the rocks must be at least 200°C. This is the temperature about 6 km underground in Cornwall. It is the shallowest depth in Britain where the rocks are at 200°C. This explains why the site was chosen. The rocks are hotter than other rocks at the same depth because the granite near Rosemanowes has higher concentrations of the radioactive elements uranium-238, thorium-232 and potassium-40. The radioactivity from these elements produces heat. This is added to the general flow of heat from the earth's interior.

◀ **Figure 4**
Volcanoes, like this one in Hawaii, provide evidence for the high temperature of the molten rocks inside the earth.

Figure 5 *Hot springs like these are being used in Iceland and New Zealand for generating electricity and for central heating. This is one form of geothermal energy.*

Figure 6 ▶
The drilling wells at Rosemanowes (far right) and venting the hot water produced by the project (right).

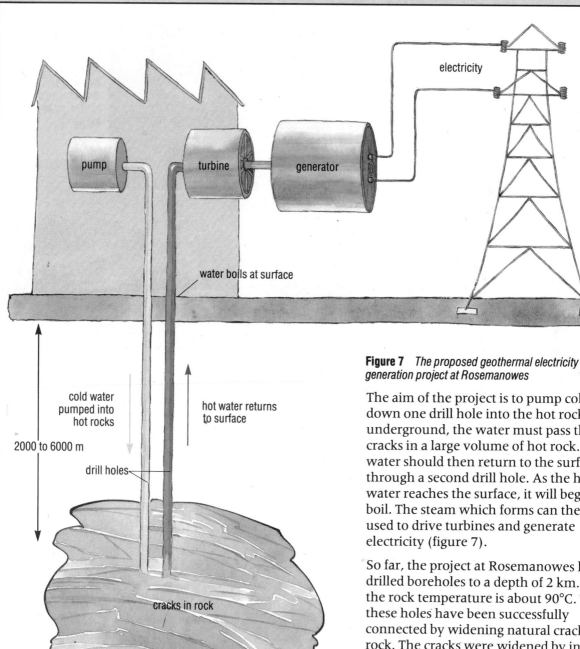

electricity

pump

turbine

generator

water boils at surface

cold water
pumped into
hot rocks

2000 to 6000 m

drill holes

hot water returns
to surface

cracks in rock

Figure 7 *The proposed geothermal electricity
generation project at Rosemanowes*

The aim of the project is to pump cold water
down one drill hole into the hot rocks. Once
underground, the water must pass through
cracks in a large volume of hot rock. The hot
water should then return to the surface
through a second drill hole. As the hot
water reaches the surface, it will begin to
boil. The steam which forms can then be
used to drive turbines and generate
electricity (figure 7).

So far, the project at Rosemanowes has
drilled boreholes to a depth of 2 km. Here
the rock temperature is about 90°C. Two of
these holes have been successfully
connected by widening natural cracks in the
rock. The cracks were widened by injecting
a gel, like wallpaper paste, at very high
pressure.

When cold water at 25°C was injected into
the rocks, the initial temperature of the
returning water was 70°C.

The project at Rosemanowes is now trying
to extract heat from rocks 6 km below the
ground at 200°C. The target is to supply hot
water to run a geothermal power station of
about 6 megawatts electrical output from
one pair of holes. If this works, several
power stations could be built. These could
supply electricity to the South-West of
England well into the twenty-first century.

1 Plan an experiment to compare the heat given up to cold
water by 1 g of hot granite with that given up by 1 g of hot
limestone.
2 Is geothermal energy renewable or non-renewable? Explain
your answer.
3 What advantages do geothermal power stations have over
a) coal-fired power stations,
b) nuclear power stations?

One of the most important issues in the world today is whether nuclear power should be used to produce electricity. In Sweden, there is a suggestion to dismantle the existing nuclear power stations. On the other hand, in France over 70% of the country's electricity now comes from nuclear power and this proportion may increase. But what is the nuclear fuel which is used in these power stations?

The important element in nuclear fuel is **uranium**. This occurs naturally in two main ores:

> **pitchblende** (impure uranium sulphide) and
> **uranite** (impure uranium oxide)

Large supplies of uranium ores come from Australia, Canada, Namibia, South Africa and the USA. It is not known how much uranium is mined in the USSR.

Enriched uranium
The atoms in uranium are not all identical. This element consists of two different isotopes – uranium-235 and uranium-238 (see page 213). Atoms of the two isotopes behave very differently in one important respect. Only the uranium-235 isotope undergoes the **fission process** which produces energy (see page 214). In the uranium ore only 0.7% is uranium-235, the rest is uranium-238. Most reactors are uranium in which the proportion of uranium-235 has been increased to about 3%. This is called **enriched uranium**.

To obtain enriched uranium, the uranium ore is first converted to uranium hexafluoride (UF_6). Uranium hexafluoride is normally a solid. For every 1000 molecules in the solid, 7 will contain U-235 and 993 will contain U-238 (figure 9).

Somehow, the proportion has to be changed to that shown in figure 10.

$^{238}UF_6$ cannot be converted to $^{235}UF_6$ by a chemical reaction because as far as chemical reactions are concerned, the isotopes are alike! Somehow, some of the $^{238}UF_6$ molecules must be removed from the mixture leaving a higher proportion of $^{235}UF_6$. This is done by heating the solid to form a gas and then using a centrifuge.

Figure 8 *Uranium ore is mined by both open-cast methods – as shown here in Gabon – and underground techniques. In open-cast mining the top soil is first removed and then the exposed uranium ore is taken away for processing.*

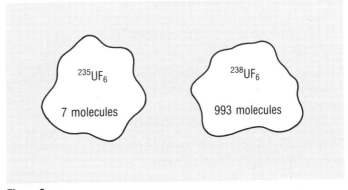

$^{235}UF_6$

7 molecules

$^{238}UF_6$

993 molecules

Figure 9

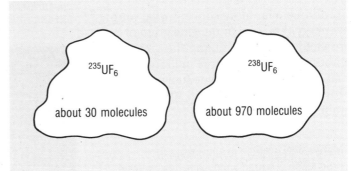

$^{235}UF_6$

about 30 molecules

$^{238}UF_6$

about 970 molecules

Figure 10

The gas passes through a tube which is spinning very rapidly, like a spin-dryer. A spin-dryer spins at about 1000 revolutions per minute. A gas centrifuge spins about 60 times faster or more (figure 11). In a spin dryer, heavier clothes are thrown towards the wall of the spinning drum in the dryer. In the same way slightly heavier $^{238}UF_6$ molecules tend to move towards the walls of the centrifuge. The gas near the centre of the centrifuge contains a slightly higher percentage of the lighter $^{235}UF_6$ molecules (figure 12). There is also a small temperature difference between the top of the centrifuge and the bottom because the motor at the bottom gets hot. So the lighter molecules also tend to move upwards.

By removing the gas near the centre and at the top of the centrifuge, chemists can obtain UF_6 with a slightly higher proportion of $^{235}UF_6$.

This slightly enriched UF_6 is passed into another centrifuge and the process is repeated. After repeating the process in thousands of centrifuges, the UF_6 gas contains about 30 molecules of $^{235}UF_6$ for every 970 of $^{238}UF_6$. The gas is cooled to reform the solid which is then converted to uranium dioxide (UO_2). The oxide will contain 30 molecules of $^{235}UO_2$ for every 970 of $^{238}UO_2$. It is this compound which is used in nuclear reactors.

The uranium dioxide is pressed into solid pellets which can be loaded into narrow steel tubes called **pins** (figure 13). The pins are between 1.5 m and 3.5 m in length.

The pins are arranged side-by-side in large steel cylinders called **fuel elements** (figure 14). Fuel elements may contain up to 200 pins. This makes it easier to load and unload the reactor (figure 15).

Uranium is a far more concentrated source of energy than coal or oil. One gram of enriched uranium contains the same amount of energy as three million grams of coal. One uranium dioxide fuel pin can provide as much energy as 150 million grams of coal.

Make a flow diagram to summarize the production of a uranium dioxide fuel pin from uranium ore.

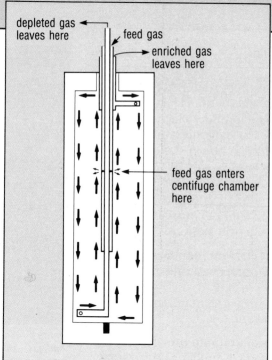

depleted gas leaves here

feed gas

enriched gas leaves here

feed gas enters centifuge chamber here

◄ **Figure 11**
As the gas centrifuge spins, heavier $^{238}UF_6$ molecules move out to the sides. The heat generated by the motor also causes the lighter $^{235}UF_6$ molecules to rise. The enriched gas is collected from the upper middle part of the centrifuge.

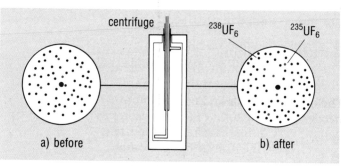

centrifuge

$^{238}UF_6$

$^{235}UF_6$

a) before

b) after

Figure 12   *The distribution of $^{238}UF_6$ and $^{235}UF_6$ molecules in a centrifuge a) before spinning and b) during spinning when there is a concentration of $^{235}UF_6$ molecules towards the centre.*

Figure 13   *Examining the quality of a fuel pin prior to loading into a fuel element* ►

Figure 15   *The reactor cap showing the fuel element entry points* ▼

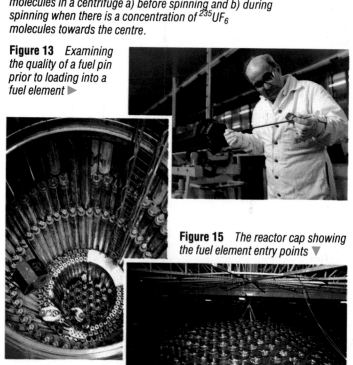

Figure 14   *Loading the fuel elements into the core tank of the reactor* ►

3 Nuclear reactors

The fuel used in nuclear reactors is described on pages 203–4. But how is the fuel used to produce electricity? Electricity is generated by turning the blades of a turbine which is connected to a generator (figures 16 – 18). The energy required to turn the turbine blades is obtained as heat energy in several ways. In a geothermal power station (see pages 201–2), the heat energy, in the form of steam, is obtained from hot rocks in the earth. In a coal- or oil-fired power station, the heat energy is obtained by burning fossil fuels which convert water into steam.

The only difference between these power stations is the method used to generate the heat to convert water into steam to drive the turbines.

The essential materials in nuclear reactors are

◆ a sufficient amount of U-235 and
◆ a source of neutrons to start the fission process.

The other materials and equipment are needed

◆ to control the process,
◆ to protect workers and the environment and
◆ to convert the nuclear energy safely into electricity.

Certain features are common to all reactors. These are described in the following five paragraphs.

a) In all reactors, the fuel elements are surrounded by a material called a **moderator**. The moderator is usually graphite (carbon), but in pressurized water reactors (PWRs) the moderator is water (figure 19).
The moderator is needed to slow down the fast moving neutrons produced by fission. Slow moving neutrons are more likely to cause fission than fast moving neutrons.
b) Further control of the reactor is achieved by **control rods**. The control rods are made of a material which absorbs neutrons. When the control rods are inserted into the reactor core, they absorb neutrons and the reaction slows

◀ **Figure 16** *The generator house showing three turbines in the background*

Figure 17 *Turbine blades being inspected prior to fitting in a generator*

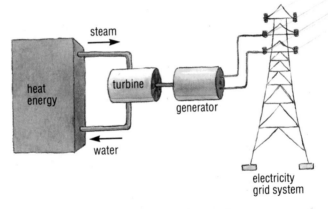

Figure 18 *Whichever fuel is used in a power station – either coal, oil, or uranium – the object is to provide heat which converts water to steam and drives the turbines.*

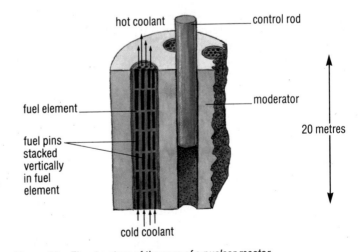

Figure 19 *The structure of the core of a nuclear reactor*

down. When the control rods are removed from the core, there are more neutrons to cause fission and the reactor heats up.

c) The heat produced by the fission of U-235 in the fuel elements is removed by a suitable **coolant**. In Britain, most reactors are gas-cooled. These include advanced gas-cooled reactors (AGRs) and Magnox reactors. The gas used is carbon dioxide.

In the USA, the USSR and many other countries, most reactors are water-cooled. The best known type of water-cooled reactor is the pressurized water reactor (PWR).

In PWRs, water acts as both the moderator and the coolant (figure 20). There is *no* graphite present.

d) Coolant gas or water passes over the fuel elements and leaves from the top of the reactor. It then passes through a **heat exchanger**. Here its heat is transferred to water in coiled tubing. The water in this second circuit boils to produce steam. The steam is then used to drive turbines and generate electricity (figure 21).

e) The reactor is contained inside a thick steel vessel surrounded by concrete **shielding**. This acts as a protection against the intense radiation and the high temperatures and pressures.

---

1 How do a) the moderator, b) the control rods and c) the coolant help a nuclear reactor to run safely and steadily?

2 How is the heat from the fission of U-235 used to generate electricity?

3 Why are reactors enclosed in thick steel and surrounded with concrete?

Chernobyl

Many of the nuclear power stations in the USSR are known as the RBMK type. It was one of these that blew up at Chernobyl (see figure 22). This type of reactor is only used in the USSR.

The moderator is graphite (carbon). The coolant is water which is pumped through the core in steel tubes. On the night of 25-26 April 1986, scientists and engineers at

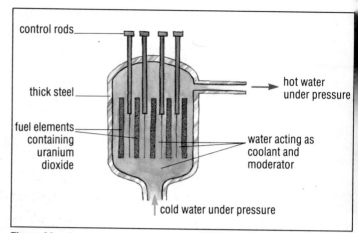

Figure 20 *A pressurized water reactor (PWR). The hot water which emerges from the reactor is used to produce steam in a second circuit. This steam is then used to generate electricity.*

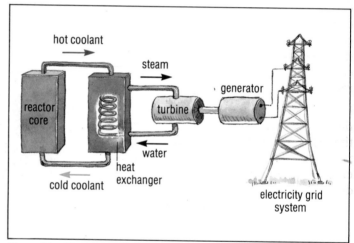

Figure 21 *The conversion of nuclear energy to heat and then to electricity in a nuclear reactor*

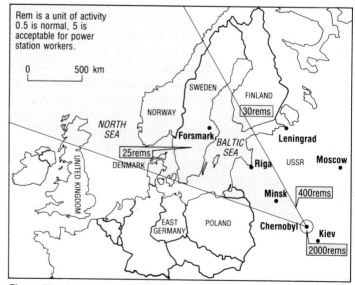

Figure 22 *Areas most affected by radioactive fallout*

Figure 23 *Plant and animal life near nuclear reactors are regularly removed for testing to monitor for radioactive contamination.*

the Chernobyl power station tried some experiments without taking adequate safety precautions. The reactor core overheated to a very high temperature causing the steel tubes to crack. At very high temperatures, carbon reacts with steam. The products are carbon monoxide and hydrogen. These gases mixed with air and exploded destroying a large part of the reactor and contaminating the surrounding area with radioactivity (figure 23).

Fortunately British nuclear power stations do not use this design. There is no chance that hot carbon and steam can meet in our reactors. Nevertheless, our scientists and engineers have learnt much from the dreadful accident at Chernobyl.

Figure 24 *The wrecked nuclear reactor at Chernobyl*

In brief
Energy today and tomorrow

1 Energy is vital to life. Energy is needed in our homes and in industry. Energy tends to spread out and become dispersed. The most useful energy sources are those in which it is concentrated. Fossil fuels (coal, oil and natural gas) provide a concentrated source of energy.

2 The earth's reserves of fossil fuels are limited. They will not last forever (figure 26).

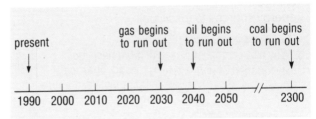

Figure 26 *Projected dates by which the earth's reserves of fossil fuels will run out*

Because of this it is important to conserve fossil fuels and prevent waste of energy. This has led to

◆ the search for alternative energy sources to fossil fuels,
◆ better methods and materials for insulation,
◆ the development of fuels for special purposes,
◆ more efficient methods of burning fuels.

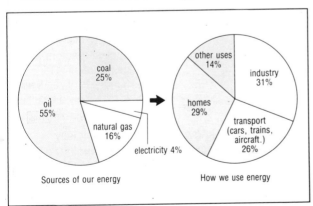

Figure 25

3 When changes occur and jobs are done, energy is converted from one form to another. Figure 27 shows the energy changes when electricity is generated from coal.

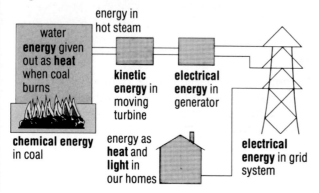

Figure 27 *Energy changes when electricity is generated from coal*

4 Fossil fuels were formed millions of years ago from decaying plants and animals. These changes involved energy conversions.

5 The process by which plants trap light from the sun and grow is called **photosynthesis**. During photosynthesis, light energy is converted into chemical energy. Diffuse ('spread out') light energy is trapped by plants and converted into concentrated ('locked up') energy in chemicals.

Figure 28 *During photosynthesis, the energy from the sun is used to convert carbon dioxide and water to oxygen and carbohydrate.* ▶

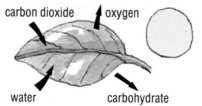

6 Material from plants and animals (such as compost, manure) can be converted into fuel **biomass**. Biomass can be used as alternative energy sources to fossil fuels.

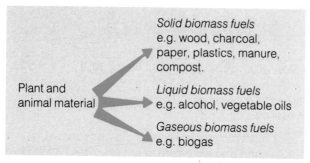

Biogas is produced when bacteria decompose organic matter (plant and animal materials) in the absence of air. This is called **fermentation**. The fuel in biogas is methane. Biogas contains about 60% methane and 40% carbon dioxide.

7 Fossil fuels are non-renewable energy sources. Biomass is a renewable energy source. Other sources of renewable energy are solar power, tidal power, wind power and wave power. Geothermal power is also considered to be a renewable energy source. Renewable energy sources are like a wage or a salary from a job – they are being replaced all the time (figure 29a). They offer great potential for the future. Non-renewable energy sources are like savings. Once they are used, they are gone forever (figure 29b).

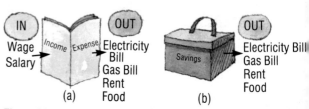

Figure 29

The most important non-renewable energy sources are fossil fuels and nuclear fuels.

8 Nuclear energy is produced from reactions in the nuclei of atoms. The nucleus of an atom contains protons and neutrons. Electrons move around outside the nucleus.

Atoms with large, unstable nuclei (like uranium) can be split by bombardment with neutrons. This splitting is called fission. Large amounts of energy are released when fission occurs.

9 There are both benefits and drawbacks to nuclear power.

Benefits	*Drawbacks*
a) Nuclear fuels will last for another 2000 years.	a) Nuclear power stations involve risks from accidents with radioactive substances.
b) Nuclear fuels produce no smoke, no soot and no acid rain.	b) There are dangers from leaks of radiation from nuclear power stations.
	c) Some of the waste from nuclear reactors will be radioactive for hundreds of years and is extremely difficult to dispose of safely.

It is difficult to know whether it is financially cheaper to produce electricity in a nuclear power station or in a coal-fired power station. Nuclear fuel is cheaper than coal. But nuclear power stations are usually more expensive to build. Getting rid of nuclear waste is also very expensive.

1 How did fossil fuels form?

Fossil fuels formed millions of years ago from dead and decaying plants and animals. 200 million years ago, the earth was warmer than it is today. Huge trees and giant ferns grew on swampy land. The sea was inhabited by vast numbers of tiny shell creatures called crustaceans. As these plants and animals died, large amounts of decaying material began to pile up (figure 30).

Figure 30 *Layers below a prehistoric forest*

Where the decaying material was in contact with air, it rotted away completely. During this process, complex compounds containing carbon, hydrogen, oxygen and nitrogen in the rotting vegetation reacted with oxygen in the air. The products of this decomposition process were carbon dioxide, water and nitrogen.

carbon,
hydrogen, + oxygen → carbon + water + nitrogen
nitrogen,   in the air   dioxide
and oxygen
in decaying
  remains

In other areas, the decaying remains were covered by sediment from rivers or by rocks from earth movements (figure 31). Here, the material decayed in the absence of any oxygen. But, it was attacked by

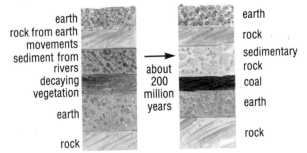

Figure 31

bacteria and compressed by the sediments and rocks above. Over millions of years, this led to the slow formation of coal from decaying plants and to the formation of crude oil and natural gas from sea creatures.

The formation of fossil fuels such as coal, oil and natural gas, involves energy changes (figure 32). Energy from the sun is trapped by photosynthesis in plants. It is converted to chemical energy in plants and animals and eventually to chemical energy in the fossil fuels.

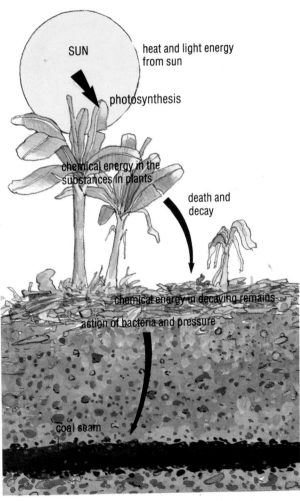

Figure 32 *Energy changes in the formation of fossil fuels*

Fossil fuels are vital sources of energy and important sources of chemicals. Many of the chemicals essential for plastics, paints, dyes and medicines are made from the constituents in oil and natural gas. When these fossil fuels become in short supply, coal will be used increasingly to make the chemicals.

Taking it further

When you see pictures of oil gushing out of the earth it is easy to get the impression that there are vast lakes of oil trapped below. But this is not so! The dark treacly liquid is absorbed in rocks, rather like water in a sponge. Rocks such as sandstone and limestone have many pores which trapped gas and oil as they were formed. These rocks lie between layers of rocks which do not have pores (figure 33). In the porous rock there is a top layer of gas, a middle layer of oil and then a layer of water. The gas is under high pressure. When the rock is drilled, the gas can push the oil up. When the gas pressure is reduced, natural gas is pumped down to push up more oil.

Some fields, such as those in the North Sea, contain natural gas and no oil.

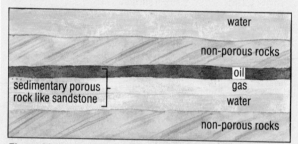

Figure 33

2 Plants and energy

All living things need energy. Animals obtain their energy by eating other animals or plants. In turn, plants obtain their energy from sunlight. Because they need sunlight, most plants grow better in the open than in the shade.

Of course, plants also need a supply of food to help them grow. The most important plant 'foods' are carbon dioxide and water. Plants absorb the energy in sunlight and then use it to convert carbon dioxide and water into glucose. This process is known as **photosynthesis**.

carbon + water + energy from → glucose + oxygen
dioxide   sunlight

Glucose is a carbohydrate, like the sugar we use in cooking. When it is heated in air, it burns to form carbon dioxide and water (figure 34). At the same time it gives out energy as heat and light.

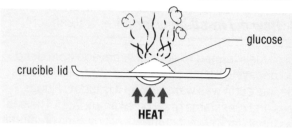

Figure 34

An equation for this reaction is:

glucose + oxygen → carbon + water + energy
                              dioxide

Notice that this is the reverse of photosynthesis.

In order to show that plants use the energy from sunlight we can compare what happens to plants kept in the light and in the dark. Plants in the light use up carbon dioxide from their surroundings and give off oxygen. Plants in the dark do the reverse and then die.

In order to photosynthesize, plants need a compound called **chlorophyll**. This is the green pigment in leaves. We can obtain a solution of chlorophyll in the laboratory by grinding some grass in ethanol (figure 35). The ethanol acts as a solvent for the chlorophyll.

Figure 35 *a) Chlorophyll in leaf cells and b) extracting chlorophyll from grass*

Chlorophyll looks green because it absorbs red and blue light from white light and reflects green light.

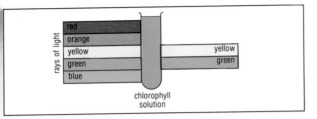

Figure 36 shows what happens when the solution of chlorphyll is placed in the path of different colours of light.

What happens to the energy from red and blue light after it has been absorbed by chlorophyll? In the test tube, it is simply radiated out again. But in plants, the chlorophyll molecule transfers the energy to other molecules so that they become more reactive. These reactive molecules can then take part in the reactions to form glucose from carbon dioxide and water.

After glucose has been formed during photosynthesis, it is used for two main purposes: as a fuel and as a building material.

Glucose molecules can link together, in plants, to form two different polymers. One polymer is called **cellulose** and the other is **starch**. Like glucose, they are also carbohydrates.

We will discuss cellulose first. This is used as a building material. In cellulose, glucose molecules are linked into long chains (figure 37).

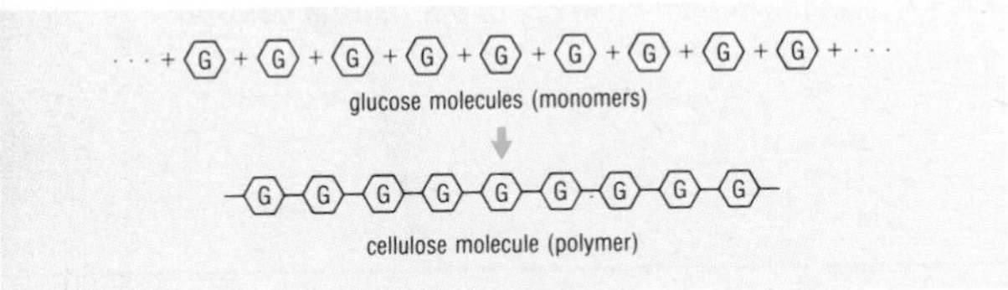

Figure 37 *Cellulose molecule (polymer)*

The long, thin structure of cellulose plays an important part in supporting plants.

In order to make starch, glucose molecules are linked in a different way. This time the polymer is branched (figure 38).

Figure 38 *Starch molecule (polymer)*

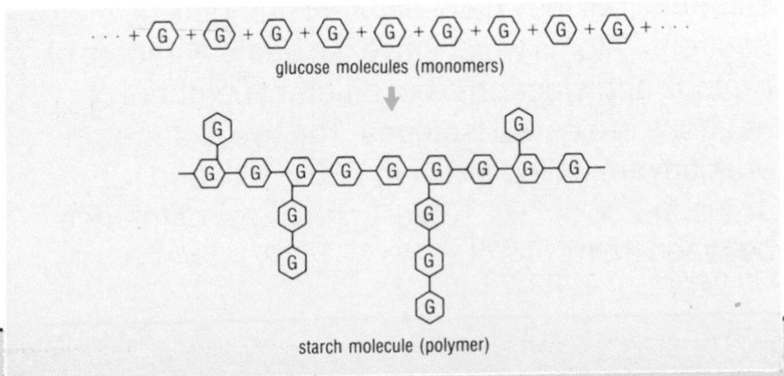

Starch is used by plants as a store of fuel. When energy is needed, the starch is broken down into glucose molecules again which are then oxidised to carbon dioxide, water and energy.

By forming starch, plants can store the energy in sunlight and then make the energy available when it is needed.

Energy is needed by plants

To grow
– which means converting glucose to starch and cellulose

To store energy
– as starch which can be used when photosynthesis cannot occur, e.g. during the germination of seeds.

Why is photosynthesis so important?

Plants photosynthesize in order to live and to grow. Animals survive by eating plants or by eating other animals which have eaten plants.

In this way, all living things rely on photosynthesis for life. We also use plant materials for a range of other purposes. These include their use as fuels. Some fuels such as wood come directly from plants. Others, such as peat and coal, are derived from plant material which grew in the past. This raises the possibility of growing plants specifically as fuels.

Figure 39 *Water hyacinths grow very rapidly under the correct conditions and will soon spread to cover the entire surface of this lake in India if left unchecked.*

The water hyacinth has been suggested as a suitable plant-fuel because it grows so quickly (figure 39). When grown on tropical sewage lagoons, it produces over 700 tonnes of dry matter per year from one hectare. Just compare this with the yield of 10 tonnes of dry matter per hectare from grassland in

Britain. These differences illustrate the factors which contribute to high growth-rate in plants. Tropical sewage lagoons have:

- high concentrations of plant nutrients (especially compounds containing nitrogen and phosphorus),
- plenty of water,
- tropical temperatures and
- high levels of solar radiation.

Even with these advantages, most plants will not grow as fast as the water hyacinth. This shows that, if we want to grow plants as fuel, we must think carefully about the conditions needed for plants to make the best use of solar energy as well as the particular plants to grow.

Taking it further

We have been thinking about energy today, even when discussing nuclear power. What about energy tomorrow?

The secret lies in harnessing the energy of the sun more effectively.

It has been estimated that, at present, the world needs 3×10^{20} J of energy each year. Written out in full, that is

300 000 000 000 000 000 000 J

The sun provides about 10 000 times more than this amount of energy each year (figure 40).

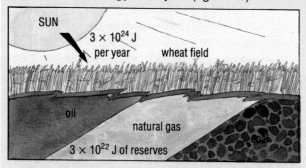

So all our energy requirements could be met if we could make solar energy converters 10% efficient and cover only 0.1% (one-thousandth) of the earth's surface with them. The problem for energy tomorrow is that we need some really effective ways of collecting and then storing the radiation. Water hyacinths and sugar cane grow quickly but only in tropical areas with a lot of water.

Chemists and biologists in many countries are trying to find ways of collecting and storing the sun's energy. One of their investigations involves finding out exactly how photosynthesis works in nature and then trying to imitate this with readily available chemicals. Another investigation is attempting to use the energy from the sun to convert water efficiently into hydrogen and oxygen. The hydrogen could then be stored and used as a fuel in the same way as natural gas (methane) is stored and used today.

Figure 40 *In 4 days the sun delivers as much energy to the earth as the known reserves of oil, natural gas and coal.*

3 Splitting the atom

Most atoms have a stable nucleus. The forces holding protons and neutrons together are very strong. Breaking these forces and splitting the atom would require a vast amount of energy. Because of this, the famous physicists J.J. Thomson and Ernest Rutherford, and their colleagues, doubted whether atom splitting could ever be used as a source of power.

All this changed in 1938. Otto Hahn and Fritz Strassman had been working for many years in their laboratory in Berlin. They were bombarding uranium with neutrons. Eventurally they found barium in the products. Barium, as you can see from the Periodic Table on page 12, is a much lighter element than uranium. It was Lise Meitner, who had been working

with Otto Hahn, and her nephew, Otto Frisch, who realised that atoms of uranium had been split. Hahn and Strassman were chemists. Meitner and Frisch were physicists. Many discoveries have been made with teams of scientists who have been trained in different subjects.

Uranium contains two different types of atom. Both types have the *same* number of protons and electrons but they have *different* numbers of neutrons. Atoms which have the same numbers of protons and electrons, but different numbers of neutrons are called **isotopes**. The two isotopes of uranium are called uranium-235 (^{235}U) and uranium-238 (^{238}U). Table 1 shows the difference between them.

Table 1 *Isotopes of uranium*

Isotope	Number of protons	Number of neutrons	Number of electrons
uranium-235	92	143	92
uranium-238	92	146	92

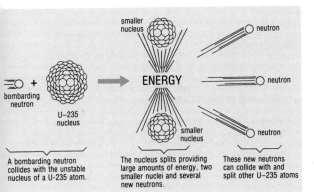

A bombarding neutron collides with the unstable nucleus of a U-235 atom.

The nucleus splits providing large amounts of energy, two smaller nuclei and several new neutrons.

These new neutrons can collide with and split other U-235 atoms

Figure 41 *Splitting an atom of U-235*

The splitting of nuclei by hitting them with neutrons is called **atomic fission** or **nuclear fission**. Notice in figure 41 that three new neutrons are released when the nucleus of uranium-235 splits. These three neutrons may then hit other U-235 nuclei producing more energy and a further 9 neutrons. These 9 neutrons may hit other U-235 nuclei producing even more energy and 27 neutrons. Given enough U-235, the process goes on and on getting bigger and bigger (figure 42). Processes like this which repeat themselves and go on and on are called **chain reactions**.

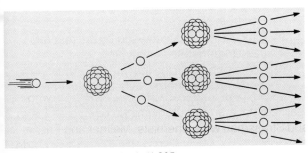

Figure 42 *A chain reaction in U-235*

How is nuclear fission controlled?

Natural uranium contains 0.7% of uranium-235 and 99.3% of uranium-238.

It is mainly the uranium-235 atoms which take part in nuclear fission. When natural uranium undergoes fission, most of the bombarding neutrons hit uranium-238 nuclei and so the chain reaction ceases. A few neutrons hit uranium-235 nuclei and the chain reaction continues in a *controlled* fashion (figure 43). This is what happens in nuclear reactors which use natural uranium or partly enriched uranium containing about 3% uranium-235.

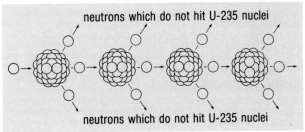

neutrons which do not hit U-235 nuclei

neutrons which do not hit U-235 nuclei

Figure 43 *A controlled chain reaction*

In an atomic bomb, fission occurs in an *uncontrolled* manner giving out enormous amounts of energy and radiation. In order for this to happen, highly enriched uranium is used. This contains a higher proportion of uranium-235. Neutrons are more likely to hit uranium-235 nuclei and an exploding chain reaction occurs. A nuclear reactor could never explode like an atomic bomb because the uranium used in the reactor is not rich enough in uranium-235.

Why does nuclear fission produce so much energy?

When a nucleus of uranium-235 splits, it forms two smaller nuclei and a few neutrons. But the total mass of the two smaller nuclei plus the neutrons is less than that of the original uranium-235 nucleus. In the process of fission, mass has been lost.

At one time, scientists believed that mass could never be destroyed. This is still true of ordinary chemical reactions in which atoms are simply rearranged. But it is not true of nuclear reactions, like nuclear fission, in which atoms of one element are changed into atoms of a different element (figure 44).

During fission, the disappearing mass is converted directly into energy. Albert Einstein explained how mass and energy were related and showed that a little bit of mass would be converted into a vast amount of energy.

Figure 44

Chemical reactions

e.g. charcoal burning

carbon + oxygen → carbon dioxide

$C + O_2 → CO_2$

Atoms only rearranged.
Mass is not lost.
Energy is produced.

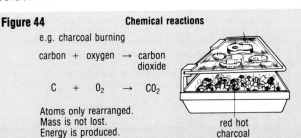

red hot charcoal

Nuclear reactions

e.g. uranium–235 undergoes fission

uranium → smaller + neutrons atoms

$U → X + Y + 3n$

New atoms form.
Mass is lost.
Vast amounts of energy is produced.

uranium bomb

Things to do

Things to try out

1 *Photosynthesis and carbon dioxide*
   A student thinks that plants cannot photosynthesize without carbon dioxide. Suggest how she could check this.
   a) Draw a diagram of the apparatus she should use.
   b) Describe the experiment.
   c) What results would you expect her to obtain?

2 *Using biogas*
   Design a biogas digestor for a small isolated village in Britain. The villagers want to use the biogas mainly for heating water and cooking at different times of the day.

Things to find out

3 Use textbooks or encyclopaedias to find out about:
   a) the uses of radioactive isotopes,
   b) the disposal of nuclear waste,
   c) nuclear fusion,
   d) the way in which Geiger counters are used to detect radiation.

Points to discuss

4 When jobs are done energy is converted from one form to another. Often there is a sequence of energy changes. What energy changes occur in the following examples?
   a) A battery is used to light a small torch.
   b) A guitar is played.
   c) Enriched uranium is used to produce electricity in a power station.
   d) A batsman hits a six and the crowd applauds.

5 Some people argue that nuclear power is dangerous and unnecessary. They say that it is a hazard to the environment and should be replaced by safer energy sources. What do you think? Give reasons for your views.

6 The British and American navies both have submarines powered by small pressurized water reactors (PWRs). Why do you think no-one has yet built a nuclear powered car?

Making decisions

7 A developing country has decided to embark on a major scheme to equip many of its villages with biogas plants for heating and cooking. What health and safety regulations do you think they should prepare for villagers operating the plants and using biogas in their homes?

8 Windmills have been used to drive machinery for hundreds of years. Recently, interest has been revived in using large windmills to generate electricity. Giant windmills with a blade diameter of 50 metres have been built on the west coasts of Scotland and Wales. Even so, one giant windmill cannot produce much electricity compared to a power station. One way to overcome this problem is to have windfarms with lots of giant windmills.

Make lists of a) the advantages and b) the disadvantages of generating electricity using windmills in a windfarm rather than using a coal-fired power station.

Questions to try

9 Read pages 205–7 carefully. Then answer the following:
   a) Explain what is meant by nuclear fission.
   b) Nuclear reactors use uranium as fuel. Give two ways in which uranium fuel differs from conventional fuels such as coal or oil.
   c) Suppose the control rods are raised a little (pulling them out of the reactor), what would be the effect on the fission reaction? Explain your answer.
   d) Why is the reactor enclosed in a thick concrete shield?
   e) Explain how heat released by the reactor is used to generate electricity.
   f) Some people argue that the UK should have more nuclear energy. Others argue that nuclear power should not be used at all. Give *two* arguments i) in favour and ii) against nuclear power.

10 This question is about the use of hydrogen as a fuel. Most fuels contain carbon. When these fuels burn, smoke and carbon dioxide are produced. When hydrogen burns no smoke or carbon dioxide are produced. Electricity can be used to produce hydrogen from water. Hydrogen has already been used as a fuel in experimental cars, but it has one major disadvantage. Even the smallest flame is sufficient to explode mixtures of hydrogen and air (oxygen).
   a) What advantage does hydrogen have compared to most other fuels?

b) Why is there no danger that hydrogen will one day be in short supply?
c) What gases would you expect in the exhaust fumes of a hydrogen-powered car?
d) Is hydrogen a perfect fuel? Explain your answer.

1 The graph in figure 45 shows the percentage of carbon dioxide in the air just above a field of grass over a period of 24 hours starting at midnight.

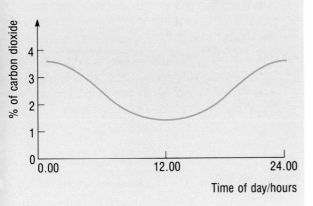

Figure 45

a) Describe how the percentage of carbon dioxide changes over the 24 hour period.
b) Why does the percentage of carbon dioxide change in this way?

c) What would you expect to happen to the percentage of oxygen just above the field of grass over the same 24 hour period?
d) Explain the predictions you made in part (c).
e) What would you expect to happen to the percentage of carbon dioxide in a busy city street over the same 24 hour period?
f) Explain the predictions you made in part (e).

12 Table 2 shows the amount of oil left and the amount being used each year in different parts of the world.

Table 2

	Amount of oil left /million tonnes	Amount being used up each year at present /million tonnes
America	18 400	800
Europe	3640	140
Middle East	61 000	610
Africa	9460	220

a) How long will the oil last in the four parts of the world listed in table 2. (Assume that it will continue to be used up at the same rate.)
b) Where is it likely that
  i) oil will last the longest,
  ii) oil will be used up first?
c) What are the major uses of oil?

If more than one page number is given, you should look up the **bold** one first.